LAB

MANAGEMENT SKILLS FOR SCIENTISTS

DYNAMICS

LAB

MANAGEMENT SKILLS FOR SCIENTISTS

DYNAMICS

Carl M. Cohen, Ph.D.

Suzanne L. Cohen, Ed.D.

COLD SPRING HARBOR LABORATORY PRESS
Cold Spring Harbor, New York

Lab Dynamics: Management Skills for Scientists

© 2005 by Carl M. Cohen and Suzanne L. Cohen
Published by Cold Spring Harbor Laboratory Press, Cold Spring Harbor, New York
Printed in India

Publisher and Acquisitions Editor	John Inglis
Development Manager	Jan Argentine
Project Coordinator	Joan Ebert
Book Designer and Production Manager	Denise Weiss
Production Editor	Rena Steuer
Desktop Editor	Susan Schaefer
Cover Designer	Michael Albano

Front cover: Background image, Photodisc Red/Getty Images. Bottom image, www. comstock.com/.

Library of Congress Cataloging-in-Publication Data

Cohen, Carl M.
  Lab dynamics : management skills for scientists / Carl M. Cohen and Suzanne L. Cohen.
    p. cm.
  Includes bibliographical references and index.
  ISBN 0-87969-741-5 (hardcover : alk. paper)
  1. Science--Management. 2. Technology--Management. 3. Scientists--Relations. 4. Organizational behavior. I. Cohen, Suzanne L. II. Title.
  Q180.55.M3C63 2005
  506'.8--dc22

                    2004030956

10 9 8 7 6 5 4 3 2 1

All Cold Spring Harbor Laboratory Press publications may be ordered directly from Cold Spring Harbor Laboratory Press, 500 Sunnyside Boulevard, Woodbury, New York 11797-2924. Phone: 1-800-843-4388 in Continental U.S. and Canada. All other locations: (516) 422-4100. FAX: (516) 422-4097. E-mail: cshpress@cshl.edu. For a complete catalog of Cold Spring Harbor Laboratory Press publications, visit our World Wide Web Site http://www.cshlpress.com/

For Zoë and Phoebe

Table of Contents

Preface

This is a book for scientists and technical professionals about surviving and succeeding in the organizations and groups in which they work. It is also for science managers and executives who want to know how best to manage scientists. If you work with scientists, this book can provide you with a better understanding of the world in which they live and the challenges that they face.

Practical advice and exercises show scientists and science managers how to interact with others in ways that improve their effectiveness and increase their productivity. The book also shows how to apply improved self-awareness and interpersonal skills to specific problems that science professionals encounter every day. If you are a scientist, the skills that you learn will enable you to better identify, focus on, and achieve your objectives. You will become more productive in your job and more successful at what you do, whether your field is molecular biology or astrophysics.

Unless you have well-developed self-awareness and interpersonal skills, all the management tools in the universe will not be of much use to you. If you are a scientist, chances are that your self-awareness and interpersonal skills are not as well developed as your technical skills; this limitation can impede your work. We provide concepts, concrete tools, and exercises that will help you to improve these skills. Our approach is designed to aid you in overcoming the barriers to knowing yourself, what to do, and how to do it.

The book draws heavily on examples and experiences from Carl's 30-year career in science, both in academia and the private sector. It also relies heavily on Suzanne's long career as a psychologist and clinician and her insights into people in general and scientists in particular. Our suggestions and guidelines work. They are all based on techniques that we have tried and used ourselves, and that we have helped others use.

A note on voice: Many of the anecdotes and experiences in the book come from Carl's career and are written in the first person. Thus, in the following, "I" refers to Carl.

Acknowledgments

We would like to express our sincere appreciation to the staff of Cold Spring Harbor Laboratory Press. Their enthusiasm, efficiency, and professionalism have made this project a pleasure from the very start. We would especially like to thank John Inglis, Executive Director of the Press, for his unflagging support and for his belief in the importance of this project. Without John's vision, this book might very well still be seeking a home.

We thank our good friend Alice Sapienza who sparked our interest in this theme and whose own book, *Managing Scientists*, was one of our inspirations. Alice also played an important role in helping to develop the workshops in which much of this book's content was first developed. We thank the American Society for Cell Biology (ASCB) and its Women in Cell Biology Committee for their belief in the importance of the themes of the book and for their sponsorship of Carl's workshops at the ASCB annual meeting for many years. We are especially thankful to Elizabeth Marincola, Executive Director of ASCB, who has been a strong supporter of our efforts and, more importantly, a close and valued friend. Indeed, the genesis of this book can be traced to a short essay entitled "Confronting the Social Context of Science," which was published in the monthly ASCB newsletter with much encouragement (and editing) by Elizabeth.

We thank Libby Koponen for her invaluable assistance in the early stages of the project, for her guidance on style, and for her snappy chapter titles as well. We also thank all of Carl's colleagues who have served as the inspiration for many of the case studies in the book. In particular, we thank Marc Charette, Simon Jones, Doug Kalish, and Per Gjorstrup for their input, support, and friendship throughout the project.

Carl M. Cohen
Suzanne L. Cohen
Newton, Massachusetts
February 2005

Introduction

The interactions of scientists in the workplace are often fascinating to the lay public because they expose the human element of endeavors that would otherwise be suitable only for textbooks. What is so riveting and at the same time paradoxical about these accounts is the immense impact of human interactions on the process and outcome of a scientific project. James Watson's account (Watson 1968) of his and Francis Crick's rivalry with Linus Pauling, their acquisition of data from and interaction with Rosalind Franklin, and their amusing and fruitful interactions with one another have been eagerly read by scientists and laymen alike. Stories about the effects of rivalries, collaborations, and relationships on the daily conduct of science, which is often imagined to be a purely rational and intellectual endeavor, are endlessly fascinating.

Despite the lively debates and discussion generated by Watson's book, little if any analysis has focused on whether the working relationships among the various protagonists could have been more productive. In this book we have tried to fill that void and, by doing so, show that there are ways to train scientists and run science organizations that improve the conduct of science, facilitate communications, and maximize productivity.

Science training programs are designed to impart technical skills and scientific knowledge to their trainees. The fact that these same programs provide no training in management or interpersonal skills sends trainees the implicit message that these skills are largely irrelevant in science. In this sense, educational institutions have two characteristics in common with most science- and technology-based organizations: (1) a single-minded focus on technology and (2) a lack of appreciation for the importance of social and interpersonal skills.

Among scientists in all fields of specialization is a strongly held belief that if you just get the science right, everything else will fall into place or become irrelevant. Yet experience shows that this belief is false. Analyses of the chemical industry (Perrow 1993), the space program (Vaughan 1996), and military campaigns (Cohen and Gooch 1991) highlight the central role that human interactions and group dynamics have in their success or failure. It has been proposed that the principal reason military commanders fail to learn from military disasters is the tendency of analysts to focus exclusively on technical and logistical explanations for failure (Cohen and Gooch 1991). This narrow focus betrays a naïve indifference to the importance of human interactions and communication and to the individual and organizational characteristics that foster them.

Efforts to train scientists and science managers to function beyond the lab bench often focus on project management, running meetings, doing performance reviews, and team building. Although scientists are efficient at learning the nuts and bolts of management—Gantt charts, work plans, etc.—mastery of these skills cannot compensate for poor self-awareness and a paucity of empathy. Without these and other personal and interpersonal skills, managerial functions will be implemented in a mechanical fashion, without heed to individual, interpersonal, or group nuances. Managing in this way makes no more sense than driving a car blindfolded: You may know how to manipulate all the mechanical controls and levers, but you are dangerously blind to context and feedback.

Interpersonal skills can be taught. For some, the abilities to relate productively, to notice how others respond to you, and to forge both personal and professional bonds come naturally. For others, and this includes many science and technical professionals, these skills are not an integral part of their personalities. Fortunately, you do not need to change your personality to become interpersonally savvy.

Interpersonal skills comprise a set of behaviors and responses that can be learned. Such learning can and should be part of the professional education of scientists. In the chapters that follow, I provide examples from my own career, illustrating how adopting some of these skills has influenced my development as a scientist, mentor, and leader.

My experience with the workshops that I have run shows that scientists are eager to learn the skills presented in this book. I have also found that the best way to learn them is to try them, and use them. Despite the fact that many scientists are initially reluctant to test new skills in role-playing exercises, their experimental nature quickly takes over. Most are ultimately convinced by the data—their own improvement as negotiators during the course of the workshop. For scientists, data rule!

It takes practice and skill to be able to observe yourself, capture what you experience, and view your behavior, body language, and facial expressions as others do. It also takes skill to attend to how others act and react to you. These skills include new ways of listening—in a manner of speaking, listening between the lines. The first section of this book describes how to acquire these skills or to improve on those you already have.

The first three chapters of the book provide a set of core skills and concepts that serve as the foundation for much of what follows. Chapter 1 provides you with the opportunity to examine your behavior in the scientific workplace in light of what is known about scientists in general. Two self-assessment exercises help you to discover which facets of self-awareness and interpersonal skills that you need to work on. Chapter 2 offers guidelines and exercises for developing or improving these skills. We teach you to become an active observer of yourself and others, and how that informs you of your feelings, helps you to choose appropriate behaviors, and enables you to assess the effectiveness of new behaviors. Chapter 3 shows how to apply your skills of self-observation, self-management, and observation in the context of negotiation. This chapter presents a framework for using these skills to guide you through the difficult situations you encounter every day. We also show you that learning and practicing negotiation is one of the best ways to acquire and improve your interpersonal skills and put them into practice.

The second section of the book teaches you to apply your new skills and powers of observation to three interpersonal domains: with employees, with peers, and with bosses. Chapter 4 offers methods for improving your effectiveness as a manager or leader of other scientists, and alerts you to the most common problems when managing teams of scientists. Chapter 5 shows you how to use these same skills when dealing with your boss. There we illustrate the most common problems that scientists have with their bosses and the most effective ways of handling them. Finally, in Chapter 6, we show that improving your ability to recognize and deal with conflict, along with practicing the skills you learned in previous chapters, will improve your ability to interact productively with peers.

The final section of the book addresses special management problems associated with the organizations in which scientists work and study. Chapter 7 describes how trainees and mentors can use self-awareness, observation, and interpersonal skills to survive in and improve the academic training experience. Chapter 8 shows how these same skills, as well as an understanding of the unique challenges that accompany the transition from academia, can improve the productivity of scientists in the private sector. Finally, Chapter 9 provides a review of the skills we have presented using an extended case study. This last chapter also suggests ways in which you can use the concepts and tools we have introduced to improve the practice of science and productivity in your own organization.

At the end of each chapter, we provide exercises and experiments designed to help you acquire and practice the skills that we present. In some cases, the exercises are in the form of experiments that allow you to use and evaluate the effectiveness of new behaviors. Use the ones that work for you.

If you believe that there is more to leading in science than intimidation, more to interacting with your peers than jockeying for advantage, and more to working with your boss than being defensive, you will find this book valuable. If you believe this but haven't a clue as to how to change your behavior, read on. The skills that you will learn will enable you to better identify, focus on, and achieve your objectives, making you more productive and effective. Finally, and perhaps most important, you will learn to accomplish all of this in a way that promotes openness and trust in your interactions with others.

REFERENCES

Cohen E.A. and Gooch J. 1991. *Military Misfortunes. The Anatomy of Failure in War*. Vintage/Random House, New York.

Perrow C. 1993. *Normal Accidents: Living with High-risk Technologies*, 3rd edition. Princeton University Press, New Jersey.

Vaughan D. 1996. *The Challenger Launch Decision: Risky Technology, Culture, and Deviance at NASA*. University of Chicago Press, Illinois.

Watson J.D. 1968. *The Double Helix: A Personal Account of the Discovery of the Structure of DNA*. Atheneum, New York.

CHAPTER 1

People Who Do Science: Who They Are and Who They Can Be

In terms of behavior patterns, affect and even some intellectual matters, we know more about alcoholics, Christians and criminals than we do about the psychology of the scientist.

MAHONEY (1979)

We are all familiar with the stereotypes of scientists portrayed in movies, television, and paperback thrillers as aloof, arrogant, intense, and distracted. Each of us is, of course, much more complex and nuanced than such simplistic characterizations, but like most stereotypes, these all have a kernel of truth in them. Scientists as a group do have personality characteristics that distinguish them from, say, social workers as a group. Although you may not share every one of these characteristics or the others discussed below, you likely share some of them. Much of this book focuses on helping you to discover which of these characteristics you share and anticipating how they may affect your behavior as well as your effectiveness as a scientist.

Many of the exercises that we present at the end of this and subsequent chapters focus on helping you to improve your interpersonal awareness and self-awareness. We focus on these two characteristics in particular because we know from personal experience that these represent areas in which many scientists are weak. We have also supplemented our personal experiences with a review of the psychological literature pertaining to the personality characteristics of scientists. Our objective in presenting this information is to help you notice and identify in yourself some of the traits that have been noted in others. As you read the following sections, take note of those characteristics that sound or feel familiar, or that others may have noted that you display. At the end of the chapter, we provide a brief questionnaire to help you to identify some of these characteristics of which you may not already be aware.

TECHNICAL PROFESSIONALS ARE DIFFERENT

The Tea Bag Company exercise

Applying the stereotypes mentioned above to all scientists seems, and certainly is, both crude and extreme. But before we dismiss such categorization out of hand, perhaps we should ask just how much truth there is to these and other popular notions of scientists as a group. The following story recounts how I first became convinced of the presence of many of these characteristics in myself.

A number of years ago, when I first became interested in improving my management skills, I took a course at the Harvard University Extension School. The course was entitled "Understanding Your Management Style," and it was taught by Robert Benfari, who had written a book of the same title (Benfari 1991). The course was intended to give students some insight, from a psychological perspective, into how they approached the tasks inherent in managing. I enrolled in this course largely because nothing else about management was offered in a time slot that was convenient for me.

During the first couple of classes, Dr. Benfari spoke at some length about "personality types" and the utility of assigning people into one of 16 categories using the Myers-Briggs Type Indicator (MBTI) inventory, a widely used and much studied psychological instrument (Briggs Myers and Myers 1995). In fact, we had taken this inventory during the first class, but had not seen the results yet. Being a scientist and a skeptic, I spent a lot of time arguing that such categorization was artificial, simplistic, and without validity.

Our third meeting consisted of an in-class exercise. Dr. Benfari read our names from a list that he had prepared, and in so doing divided us into four or five groups. He told us that within our groups we were to come up with a plan to address and solve an organizational problem that he would pose for us. He had also assigned one student from each group to observe how the group went about its task, without participating. He referred to the exercise as the "Tea Bag Company" exercise.

Each group was to put itself in the role of senior managers of Tea Bag Company, Inc. As managers, we had just learned from our sales and marketing department that our sales were down catastrophically over the last two quarters. Our task was to determine a solution to this problem. We would have 45 minutes to work on the problem within our groups, and then each group would report to the class on what they had decided to do.

Within my group, several members immediately suggested that we convene near the white board so that we could use it to organize our strategy. We did so, effectively preventing any other group from accessing the board. Within a few minutes, my group was intensely involved. Members were interrupting each other, talking in loud voices, and grabbing the marker from one another to write on the board. We all agreed that we needed to take an analytical approach to the problem. We mapped out a marketing survey to determine whether consumers' tastes had changed. We allocated resources for analytical testing of our tea bags to see if quality had slipped. We crafted a backup plan to move into coffee, if that seemed prudent. We were really very efficient and logical, and completed our task easily within the allotted time.

After the 45 minutes, Dr. Benfari reconvened the class, instructing us to report one group at a time. My group was asked to report first. One of us outlined the series of logical steps we took and how we focused on objective measures of success and economic outcomes. The observer accurately described our deliberations as being lively and competitive, and noted that we all jockeyed for board time, interrupted one another, and spoke more than we listened. This was not suprising to our group; it was how we behaved all the time.

Dr. Benfari then asked the second group to report. The spokesperson for that group said that the group's primary concern was the welfare of the employees of the Tea Bag Company. The group believed that since the problem was so acute, a plan should be in place to ensure that if the company were to go under, the employees would be provided for; they would have adequate outplacement services, and health benefits would continue for as long as possible. They also scheduled an emergency stockholders meeting to allay the concerns of the company's investors. They did start to deal with how to address and fix the sales problem, but had not progressed very far when their time ran out.

I recall listening to this presentation and thinking that these people must have landed in corporate America from the Moon. I was baffled by their approach. I was further baffled when the observer assigned to that group reported that the discussion had been quiet and respectful. The observer said that group members waited for one another to finish speaking before speaking themselves, and that one member of the group had gone out and brought back sodas for the whole group during their discussion.

After the other groups reported, it became clear that a very wide spectrum of approaches had been taken. But none was so remarkably different from my group as that of the second group that had reported.

When all of the reports had been delivered, Professor Benfari told us that he had composed the groups using our personality types as determined from the MBTI inventory that we had taken during the first class. He said that my group was dominated by "NTJ" personalities, which in Myers Briggs jargon stands for intuitive, thinking, and judging. We do not need to go into the arcana of the Myers Briggs categories; suffice it to say that NTJ people are the typical scientist types (for a more detailed discussion of MBTI types and how they relate to a chosen profession, see Tieger and Barron-Tieger 1992), who have a tendency to be highly intuitive (N), analytical and logical (thinking: T), and can be very judgmental (J).

The second group in the class was composed largely of members determined to be "NFP"—or close to it. NFP stands for intuitive (N), feeling (F), and perceiving (P). NFP people are highly relational and outgoing, and use their feelings to come to conclusions (intuitive). They react to the world in a feeling mode, rather than a thinking or analytical one. In other words, they are in many ways the exact opposite of the NTJs.

I was completely dumbstruck by this revelation. None of us had known the basis on which we were placed into our respective groups. We all went about working on our task in ways that came naturally to us and we behaved precisely as the MBTI would have predicted.

People really are different, and they are different in ways that can be described and measured. As much as I hate generalities and categorization, I know that I and many of my scientist colleagues are NTJs or STJs. The "S" ("sensing") suggests that some of us have a more data-driven way of coming to conclusions, compared with the Ns, who are more intuitive. And I also know that we work in ways that are different from how NFPs and many others work. I became a believer in the MBTI, not as a diagnostic or classification tool, which is how it is typically used, but as a tool for insight into myself. Do not be fooled into thinking that just because the MBTI can identify people who share behavioral characteristics that it should be relied on to choose a profession or direct others into a profession. As has been amply noted, most recently by Annie Murphy Paul in her book *The Cult of Personality: How Personality Tests Are Leading Us to Miseducate Our Children, Mismanage Our Companies, and Misunderstand Ourselves* (Paul 2004), the predictive value of these tests is overrated and the tests themselves overused.

We refer to the MBTI only for the purpose of illustrating that scientists as a group are superb at focusing on tasks, but less attuned to the interpersonal. To go about managing scientists without taking into account who they are as people, and how their personalities differ from other types of people, is like trying to train a pack of tigers using a training manual meant for parakeets.

What research shows about the personalities of scientists

The Tea Bag story suggests that scientists and technical professionals share certain behavioral characteristics. In the following section we review a few of the studies that have attempted to identify these characteristics. Despite the fact that the quote at the beginning of this chapter suggests that there is a paucity of such data, the data that exist are revealing.

The opening quote was cited in a comprehensive review of the psychology of science and scientists by Feist and Gorman (1998). This review contains references to more than 150 scholarly publications relating in one way or another to the psychological characteristics of scientific and technical professionals. The following is a list adapted from that article, and compiled from the literature of experimental psychology, that compares the personalities of scientists to those of nonscientists.

Compared to nonscientists, scientists are

- More conscientious and orderly
- More dominant, driven, or achievement oriented
- More independent and less sociable
- More emotionally stable or impulse controlled

A bit more intriguing is the summary in the same article of the differences in personality between "eminent" and "less eminent" and "creative" and "less creative" scientists (let us not obsess here over how eminence and creativity were quantified). According to the article, compared to less eminent and less creative scientists, eminent and creative scientists are more

- Dominant, arrogant, self-confident, or hostile
- Autonomous, independent, or introverted
- Driven, ambitious, or achievement oriented
- Open and flexible in thought and behavior

Beginning to get the picture? Of course, there is always the issue of cause and effect. Does a career in science promote arrogant, antisocial behavior, or does science attract those who already have a tendency to exhibit these characteristics? Feist and Gorman take the safe middle road and suggest a bidirectional interaction between personality and science.

In another publication, Greene (1976) reported that "The psychological problems most frequently encountered with...scientists stem from (a) communications difficulties, (b) confusion about the role of the expert, (c) emotional and interpersonal needs, and (d) failure experiences."

In a study of 99 academic researchers (all full professors), Feist (1994) concluded that "[eminent scientists]...were rated by observers as more exploitative, more fastidious, more deceitful, less giving, and less sensitive to the demands of others... . In sum, complex thinkers about research are influential in their discipline and are well cited, but are considered by observers to be neither warm nor sociable."

Finally, in a study of 100 technical project team leaders (in unspecified technical areas) Gemmill and Wilemon (1997) listed the top ways in which scientific and technical project team leaders misread events within project teams. These leaders

- Were unaware of interpersonal conflict among members of the team

- Were unaware of hidden agendas on the part of team members

- Did not understand the motivation and needs of team members

- Were unaware of expectations of team members

- Did not listen carefully to team discussion

- Misread lack of argument as agreement

- Interpreted conflict as unhealthy when it was actually constructive

- Misread team members' ability to work together as a team

So, at the considerable risk of overgeneralization, the data suggest that as a group, science and technical professionals are poorly attuned to the dynamics of their interactions with others and to the needs and feelings of those around them.

WHY YOU SHOULD PAY ATTENTION TO YOUR PERSONALITY

If the only consequences of exhibiting some or all of the traits mentioned above were that you might be viewed as being aloof and uncaring, you might be excused for having little or no motivation to take note of them in yourself. However, the consequences of such traits and the behaviors they engender can be far more profound, even to the point of being dire. Let us examine some hypothetical consequences in the science workplace of a few of the personality traits identified in the studies cited above. These brief vignettes outlining the consequences of behaviors, which were found to be common among scientists, are based on actual cases.

String together enough outcomes like those listed below and before you know it, the people in your group, company, or organization are confused and alienated, projects are foundering more often than they should, and decisions are being made for other than scientific reasons.

Trait	Consequence
Dominant, driven, or achievement oriented	You forge ahead on projects with your own ideas, listening politely but usually failing to take into account the suggestions or objections of colleagues or employees. Most of the time, this works well because you know more than they do. But the one time that you do not, you continue to follow your own agenda and end up spending millions of dollars on a failed project that you should have abandoned two years ago.
Arrogant or hostile	Of course you are arrogant: You are the smartest, most accomplished scientist in the company. But when it comes time to seek the support of others for a controversial new technology that you want to acquire, you find yourself isolated and without support. The technology is actually just what your company needs, but because you have created enemies with your arrogance, you cannot get anyone else to agree with you. The company suffers and so do you.
Introverted	Paying attention to other people is a distraction. It takes you away from your work. Moreover, it is hard, and you figure that people are complicated and unpredictable. You fail to notice that, over time, people are excluding you from their informal discussions because you would rather be in your office analyzing data. The result is that you do not hear about the new project until the formal announcement is made, by which time all the team leaders have been selected. You wonder why you were left out.
Less sensitive to the needs or demands of others	You figure that just like you, everyone has their own agenda, and it is pretty hard to figure out what that agenda is. You have always felt that people complain all the time—it is only natural. One day your senior and most valuable postdoc announces that she is leaving in three weeks for a job in industry where the salary is higher and the advancement prospects are better. You recall that over the past several months, she has been asking you about a salary increase and whether you would support her for a faculty position, but you kept putting her off. Now you are faced with the prospect of a major setback in your most important research project.
Unaware of interpersonal conflict among members of the team	You hired Alice and gave her the same project as Axel because you thought that competition would drive them both to work harder. The result was that Axel hoards reagents and signs up for equipment time that he does not need to prevent Alice from getting the better of him. Others in the lab mention the brewing conflict, but you shrug it off with the comment that the competition will make each of them stronger. Three months later, Alice goes to human resources and files a sexual harassment complaint against Axel. The resulting turmoil sets both of them, and the lab, back a year.
Unaware of hidden agendas on the part of team members	You are trying to decide on the appropriate version of a recombinant protein with which to go into production. The director of protein expression has been arguing vehemently for version 2C, whereas others in the group believe that several other variants are more appropriate. You believe that everyone is arguing the case for each variant on its merits. Months later, you learn that the reason the director was arguing for 2C had nothing to do with its scientific merits: He had prematurely anticipated the use of 2C and had his group produce several grams of it. Had you known this, you might have reassured him that you would have been happy to sacrifice the produced 2C in favor of making the best choice.

Trait	Consequence
Unaware of expectations of team members	As head of a task force on in-licensing a new technology for fabricating a new type of solar cell, you ask one of your scientists to review what is known about the physics behind the technique. She spends a week on the project with the expectation that she will present her conclusions to the executive committee that makes the final acquisition decision. When she delivers her findings to you, you indicate that they will simply be attached as an appendix to the final report. She accuses you of misleading her, and you counter that she spent way too much time on the report. She loses trust in you, and you are annoyed by what you see as her unreasonable expectations.
Does not listen carefully to team discussion	During a group meeting, one of your postdocs comes up with an idea for a new project. Others in the group are enthusiastic, but you are preoccupied because you are answering e-mails on your laptop. A month later, you suggest this same project to a new postdoc. When the first postdoc finds out about this, he assumes that you stole his idea intentionally and complains to the department chair. Several days of everyone's valuable time are wasted sorting out the misunderstanding.
Interpreted conflict as unhealthy when it is actually constructive	You are managing a project team of engineers and biologists developing a new brain imaging technology. The engineers keep insisting that the biologists have not collected enough data on glucose metabolism in the brain to ensure that the instrument will have adequate dynamic range. The biologists insist that the engineers are being overly compulsive and are demanding data impossible to obtain. The groups have reached an impasse. You believe that the project will self-destruct unless you can defuse the situation. You finally insist that the groups cease arguing and that the engineers move ahead with the design despite the disagreement. When tested in the clinic, the prototype instrument is not sensitive enough to changes in glucose metabolism to be useful.
Misread team members' ability to work together as a team	Your group will move into a new lab in six months and you are finalizing its layout. Lab benches and office space need to be assigned. You are about to leave for a meeting in Italy and tell your group that they can work out the assignments themselves and you will review their recommendations when you return. Several members of the group feel disenfranchised by the process, which is dominated by a few highly aggressive group members. They accept space assignments that do not meet their needs because they refuse to argue with the others. They never mention this to you, but their morale deteriorates and they minimize their interactions with the others in the lab. You never notice.

What is so insidious about the behaviors in the above examples is that in each case the protagonist was behaving in a "reasonable" manner. No overt hostility was evident and there were no actions that might be cause for allegations of misconduct or mistreatment. Moreover, in many of the examples the actions of the protagonist were the result of considerable thought and deliberation. If all of this is true, what went wrong? Quite simply, the thought and deliberation were all focused on scientific and technical matters and not at all on interpersonal consequences.

Whereas the above examples were heavily weighted toward team leaders as protagonists, bench scientists and technicians are no less prone to encountering such problems. Indeed, even the most mundane interactions during a typical day in the lab can have unintended consequences for scientists who fail to anticipate the effects of their behavior on others. A typical day for a working scientist might include the following interactions:

- Meet with technician to discuss plans for the day. Since technician also works for colleague, must negotiate technician's schedule with colleague. Colleague uncooperative.

- Talk to different colleague, request some of her confocal microscope time, since microscope schedule is fully booked for next two weeks. Colleague refuses. Stomp back to bench and wonder how to finish experiment on schedule.

- Start weighing out reagents for experiment. Discover that previous user spilled unidentified chemical on electronic balance. Instantly identify culprit and resolve to confront him.

- Schedule time to meet with lab director to discuss attending the annual meeting of the American Society for Cell Biology, although she had already said that travel funds were exhausted. You are angry because two of your peers are going.

- Start experiment, try to put irritation with colleagues out of mind, and concentrate on work. Become distracted by loud rock music coming from next lab bench. The longer you hear it, the angrier you become.

These are fairly typical, perhaps even understated, examples of interactions with which scientists deal on a daily basis. Although they take place within the context of professional activities, how you deal with them will depend on your personality. Some avoid conflict, whereas others exacerbate it; some become aggressive and others withdrawn or reticent; some act in ways that establish productive alliances with colleagues, whereas others work in isolation; and some have satisfying relationships with supervisors and others spend their careers feeling manipulated and unappreciated.

Each of these contrasting behaviors produces very different interpersonal and professional consequences. The success of your work and the progression of your career are strongly influenced by your behavior and whether you interact with others in a productive or an antagonistic manner. For example, take the examples of typical laboratory interactions that were listed above:

- If you become accusatory and confrontational when trying to negotiate regarding your technician and time on the confocal microscope, not only do you risk not getting the time you need, you may create animosities that make these negotiations more difficult in the future.

- If you accuse your colleague of spilling chemicals, you may be blaming the wrong person, or engender such an angry denial that you end up cleaning the mess yourself.

- If you go to your lab director contending that you are being unfairly treated and become accusatory or insulted, you may actually reduce your chances of changing her mind.

- If you continue to bottle up your annoyance at the loud music, you may end up losing your temper with someone else over an unrelated and trivial matter.

Of course, everyone has bad days, snaps at a colleague, or says something thoughtlessly that they later regret. But it just may be that scientists do this more often than others.

At one of my workshops for scientists, I asked participants to answer several questions about ways in which their interactions with others in the lab affected their work. Here are some of their responses to three of the questions:

- More than three-quarters reported spending between 10% and 25% of their time at work on "people problems."

- More than two-thirds reported having between one and five "uncomfortable interactions" with people at work each week.
- Nearly two-thirds reported that interpersonal conflict had hampered progress on a scientific project between one and five times during their career.

If these figures are even close to being representative of the science community as a whole, we are wasting a lot of manpower, resources, and time because of interpersonal problems. Many of the scientists in my workshops routinely ignore these problems or try to resolve them in ways that create more animosity than was present to begin with. It is not uncommon for us to avoid problems in the workplace because we lack the skills to resolve them diplomatically. But scientists as a group, and science organizations as a whole, may be more prone to such avoidance than others. Why is this so?

During years of biomedical research, I spent countless hours regularly reading the scholarly scientific literature in my field. I also read journals dedicated to promulgating the latest time-saving and clever techniques. I even read advertisements for products that promised to accelerate my research with the latest technological innovations. I spent several weeks each year at scientific meetings to learn of the latest ideas, discoveries, and techniques that might help me in my work. At these meetings, I even went so far as to speak with the dreaded salesperson about new instrumentation, reagents, and gadgets for my lab. I did all this religiously for almost 15 years before I thought to seek out any information about managing my laboratory and its research personnel more effectively.

If scientists are willing to invest time and effort in learning a new skill, why not do the same for management and interpersonal skills? And why, when they do decide to do so, do they wait until the need is acute? Recently I was engaged by a research organization to help resolve what it referred to as a crisis among some of its senior members. Many of them were not on speaking terms with others, and the leader of the organization was at his wits' end about how to solve the problem. When I asked how long this situation had been going on, I was stunned by the answer: ten years!

Not everyone waits ten years, of course. Many people attend my workshops to proactively gain the skills that I teach, and many more attend in response to a particular problem that they are having at that time. When I ask participants for questions at various points in my workshops, I am always struck by how specific they are: "What if you had a person in your lab who always lied about whether they had made the mess in the fume hood?" or "Let's say you worked for a group vice president who was always trying to micromanage everything you did?" It is clear that these are not hypothetical examples—they represent issues regularly dealt with at work.

THE GOOD NEWS: SCIENTISTS ARE PSYCHOLOGICALLY FLEXIBLE AND QUICK LEARNERS

Fortunately, data from the scholarly studies that we cited above are not all bad news for scientists. The studies also found that scientists were emotionally stable, impulse controlled, and open and flexible in thought and behavior. What this suggests is that despite less-than-optimal interpersonal skills, technical professionals have a high capacity, motivation, and willingness to learn and improve. What they need is data showing the utility of improvement, as well as the opportunity to learn.

Technical professionals may spend as many as ten years in college and professional school yet never experience a single hour of training to help them manage themselves and others. The training in working with and managing others that they do receive comes from observing the behavior of their mentors, many of whom are themselves untrained and often poor managers (see Chapter 6 for more on this topic). If you are a scientist working for a company, chances are that you have been sent to one or more management training seminars over the course of your career. On the other hand, if you work in academia, it is likely that no one has even suggested that you attend a management training seminar.

Whether run by your company or an outside agency, these seminars typically focus on the nuts and bolts of managing: budgeting, time management, goal setting, and project management. These are all important skills and worth learning. However, your success at applying these skills will not be determined by how well you learn them or even how long you use them. Your success will be determined by how well you understand, relate to, and respond to the people you manage and with whom you work.

Standard management training may provide concrete guidance for dealing with *overt* behaviors, but it does not help you to see beneath the surface to the underlying motivations or needs that drive those behaviors. Thus, you may have learned how to fill out the annual performance review form for your employees, but if you present your feedback in an aloof or devaluing manner, you may do more harm than good. You may have learned how to create and implement a project plan, but if you fail to notice and deal with conflict among members of the project team, your plan may founder. You may have learned how to organize and run group meetings, but if you fail to notice and address the fact that several key participants routinely remain silent during these meetings, you may be running the project at half steam.

If you are oblivious to conflict, insensitive to the needs and aspirations of others, and unaware of the impact of your own behavior on others, you are managing under a handicap. If you are interpreting silence as agreement, repeated absences as laziness, or failure to follow instructions as forgetfulness, you may be missing important underlying dynamics that can hamper or derail an important project.

The good news is that it does not have to be this way. Improving your interactions with others does not require a personality transplant, and learning how to notice and manage your reactions and behaviors in difficult situations does not require years of psychotherapy. The following chapters present steps that you can take to improve your situation, and they outline ways in which scientific organizations can take the lead in promoting and legitimizing the importance of interpersonal expertise as much as they promote technical expertise.

SUMMARY

Research shows that professionals in science and technology are more likely than others to have personality characteristics that lead them to avoid or miss important interpersonal cues. They may also act in ways that show a lack of appreciation of the effects of their own behavior on others. These traits can have unanticipated negative consequences on their careers and scientific progress. Becoming aware of your own personality characteristics is the first step towards becoming a more effective scientist and science manager.

REFERENCES

Benfari R. 1991. *Understanding Your Management Style: Beyond the Myers-Briggs Type Indicators.* Lexington Books, New York.

Briggs Myers I. and Myers P.B. 1995. *Gifts Differing: Understanding Personality Type.* Davies Black, Mountain View, California.

Feist G.J. 1994. Personality and working style predictors of integrative complexity: A study of scientists' thinking about research and teaching. *J. Pers. Soc. Psychol.* **3:** 474–484.

Feist G.J. and M.E. Gorman M.E. 1998. The psychology of science: Review and integration of a nascent discipline. *J. Gen. Psychol.* **2:** 3–47.

Gemmill G. and Wilemon D. 1997. The hidden side of leadership in technical team management. In *The Human Side of Managing Technological Innovation: A Collection of Readings,* 1st edition (ed. R. Katz). Oxford University Press, New York.

Greene R.J. 1976. Psychotherapy with hard science professionals. *J. Contemp. Psychotherapy* **8:** 52–56.

Mahoney M.J. 1979. Psychology of the scientist: An evaluative review. *Soc. Stud. Sci.* **9:** 349–375.

Paul A.M. 2004. *The Cult of Personality: How Personality Tests Are Leading Us to Miseducate Our Children, Mismanage Our Companies, and Misunderstand Ourselves.* Simon & Schuster, New York.

Tieger P.D. and Barron-Tieger B. 1992. *Do What You Are: Discover the Perfect Career for You through the Secrets of Personality Type.* Little, Brown and Co., Boston, Massachusetts.

EXERCISES AND EXPERIMENTS

1 Self-assessment: Who you are

This and the following exercise are designed to help you become more aware of how you interact with others and how your behavior influences others' responses to you. It is a self-assessment in that you are answering the questions about yourself. The goal is not to label yourself as having one character type or another. Rather it is to provide a practical framework within which you can notice or observe your own behavior in the workplace. By the very process of thinking about and answering the questions, you will have taken an important first step in recognizing how your feelings, reactions, and responses can impact your effectiveness in the scientific workplace. The traits or characteristics listed on page 13 were taken from several of the studies described in the section beginning on page 4 of this chapter.

The traits in the self-assessment exercise on the following page ("Identify your traits") are separated into groups that have the following characteristics:

Group 1. May limit your effectiveness in managing or collaborating with others. May result in your being seen as untrusting and uncollaborative.

Group 2. Inherently neutral, but if manifested in an extreme manner, can result in consequences similar to those of group 1.

Group 3. Characteristic of task orientation and achievement; generally thought of as positive attributes. If traits are overly dominant, you may be seen as "cold-blooded" or self-centered.

Group 4. Contribute to the maintenance of collaborative and productive group interactions.

Read through your responses to the exercises. Does the overall picture describe you? Do you see trends that surprise or dismay you or limit your effectiveness as a leader or team member? If so, make note of them and pay special attention to sections of the following chapters that address those issues.

If you found that you had more than one or two entries checked "do not know" in the "I think I am" group, this is a hint that your self-awareness may need some work. If you do not know how to characterize your attitudes, feelings, or reactions, you are not paying close-enough attention to your own behavior. The next chapter provides exercises to help improve your self-awareness. Take this inventory again in few months, after you have worked on self-awareness.

If you checked a lot of "do not know" in the "others think I am" category, you may need to improve your skills at paying attention to how others react to you. The section beginning on page 106 of Chapter 6 may help you in this arena. Retake this inventory after you have read that section and note whether your answers have changed.

2 Self-assessment: Dealing with others

The goal of this exercise is to help you to identify situations involving others that you find difficult. Like above, the objective is not to provide a diagnosis or assessment of your

Group/traits	I think I am			Others think I am		
	Yes	No	Do not know	Yes	No	Do not know
Group 1						
Overly dominant						
Arrogant						
Hostile						
Introverted						
Uncommunicative						
Deceitful						
Ungiving						
Insensitive						
Group 2						
Autonomous						
Driven						
Ambitious						
Fastidious						
Group 3						
Conscientious						
Orderly						
Emotionally stable						
Impulse controlled						
Self-assured						
Achievement oriented						

Exercise: Identify your traits

Group/traits	I think I am			Others think I am		
	Yes	No	Do not know	Yes	No	Do not know
Group 4						
Open and flexible						
Aware of conflict						
Aware of hidden agendas						
Able to understand others' motivations and needs						
Able to listen carefully to discussion						
Able to channel conflict to achieve desired results						
Good at identifying people who are compatible with one another						

problems, but rather to help you notice and remember what you find difficult or uncomfortable when dealing with others. You may find it useful to return to these same questions periodically, every six months or so, to assess whether your views change as you develop the skills presented in the following chapters.

1. Describe in as much detail as you want one specific difficult interpersonal interaction that took place in the context of your work as a science professional. Describe the impact this interaction had on your work.

2. List three other difficult situations and specify the nature of your relationship with the person(s) involved (example: unpleasant conversation with another postdoc about authorship on a paper). In this case, write no more than one sentence for each.

3. How many times per week do you engage in what feels like an uncomfortable interpersonal interaction with someone in your scientific workplace?

___never   ___1–2 ___3–5   ___6–10   ___more than 10   ___too many to count

4. Rank the following categories of people in order of frequency of difficult or conflictual interactions over the past year (4 = most frequent; 1 = least frequent).

___colleague or peer   ___direct superior
___employee   ___administrator or clerical worker

5. Approximately how many times in your career has progress of a scientific project been negatively affected by an interpersonal conflict that was not handled well?

___never ___once or twice ___3–5 times ___more than 5 times ___do not know

6. About what percent of your time do you spend thinking about or dealing with interpersonal or "human" issues in your professional day?

___none   ___less than 10%   ___10–25%
___26–50% ___51–75%   ___more than 75%

7. Check the boxes on page 16 that best describe the degree to which you agree or disagree with the statements (scale: 5 = strongly agree; 1 = strongly disagree).

Like the previous inventory, take note of the frequency with which you answered "do not know." If you answered this way to questions about observations of yourself, pay attention to the exercises for self-awareness at the end of Chapter 2. If you answered this way to questions about how others see you, pay attention to the exercises for observing others at the end of Chapter 6.

3 Identifying themes

How easy was it for you to answer the above questions? List any questions that seemed particularly difficult, either because you had a hard time understanding or answering them. Questions that you found difficult or troublesome may refer to interpersonal themes or personal characteristics that you find uncomfortable to think or talk about. Making note of these themes may enable you to anticipate work situations in which you are uncomfortable and in which you may not perform optimally. As we show in the following chapter, such anticipation is often the key to intercepting ineffective behaviors.

Are you able to identify themes in your answers that give you better awareness of our own behavior and reactions? By answering these questions, have you identified some behavioral themes that are common to the various situations?

Some examples of themes that you might identify include

- "I have a lot of difficulty dealing with my peers"

- "There is a disconnect between the way I see myself and the way others see me"

- "I was unaware of how much my behavior affects others"

By answering these questions and finding common themes, you are becoming more aware of your interpersonal style, the first step toward managing yourself and others in a more informed manner.

	Strongly agree 5	4	3	2	Strongly disagree 1	Do not know
I am very collaborative						
Others think that I am very collaborative						
I think that I am confrontational and argumentative						
Others think that I am confrontational and argumentative						
I am sensitive to others' needs and feelings						
Others think that I am sensitive to their feelings						
I am receptive to suggestions from others						
Others see me as being open to suggestions						
I avoid interacting with colleagues that I do not like						
I manage to interact with colleagues as needed, regardless of whether I like them						
I tend to withdraw in tense or conflictual situations						
I tend to become aggressive in con-flictual situations						

CHAPTER 2

The Mote in Your Own Eye:
Manage Yourself First

Not only are scientists less likely to be attuned to the dynamics of interpersonal relationships, but they are also less likely to notice and manage their own emotions and reactions. Recognizing these characteristics in yourself is the first step to improving them.

In this chapter, we show how self-awareness can improve your ability to manage yourself and others, and improve your effectiveness at what you do. We review case studies that illustrate problems scientists typically encounter and show how improved self-awareness can help you to deal with these problems more effectively.

THE IMPORTANCE OF BEING SELF-AWARE

Some of the underlying themes of this book, especially as they relate to self-management and self-awareness, were influenced by the "emotional intelligence" school of thought. Daniel Goleman's book *Emotional Intelligence* (Goleman 1995) as well as the more recent book *Primal Leadership* (Goleman et al. 2002) argue convincingly for the application of the skills associated with emotional intelligence in the workplace and also suggest that curricula need to be developed that are aimed at teaching these skills at every level of education.

One of the key elements of emotional intelligence is self-awareness, the capacity to notice what we feel and think. The ability to do this in "real time," that is, to be able to notice and name feelings and reactions as you experience them, is the hallmark of self-awareness. Another element of self-awareness is the capacity to anticipate your own behavior, i.e., imagining what you will or could do in the moment, before you actually do it. For example, experiencing anger and anticipating that your response might be to berate or insult the person against which it is directed places you in the position of being able to decide how to behave, instead of simply behaving in accordance with how you feel. Thus, self-awareness allows you to exercise behavioral options and choose the behavior that will be the most effective, rather than the one that may make you feel good for the moment, but that you will later regret.

Let us start with an example from my own career of how this process works.

▸▸ *Case Study: Left Out*

I worked with a team on a drug discovery project that involved personnel from all parts of the company and required multiple team meetings every week. During an especially tense phase of the project, the meeting organizer told me that the weekly meeting with senior managers would not be held that week. However, when the day of the meeting arrived, I learned by accident that the meeting was in fact being held, and that several other senior personnel were planning to attend. I could not ascertain who knew about the meeting and who did not. At this point, my sensitivity to rejection took over. My first reaction was to assume that the meeting was being held without me and I was intentionally being excluded. I had been especially outspoken at the last meeting and worried that my viewpoints were so objectionable that the team decided that it would be best to leave me out of the next meeting. My reaction to this feeling was to resolve never to go to another meeting unless I was given a personal invitation. My second reaction was to mentally compose stinging e-mails of rebuke and invective to the organizers of the meeting.

I spent some time experiencing these feelings of withdrawal and hostility without acting on them. I knew from past experience that I have a particular sensitivity to being excluded, and that I had on other occasions misinterpreted other's actions as being exclusionary when in fact they had nothing whatsoever to do with me. Reminding myself of this helped me to put my current feelings in perspective and made me cautious about saying or doing anything about the situation until I had more information. Even though I could not imagine why they were having the meeting without me, I nonetheless resolved to keep an open mind, refrain from jumping to conclusions and, more importantly, avoid reacting overtly to my feelings.

The next day, I discovered that the meeting really had been cancelled but some of the participants had decided to get together at the original meeting time to discuss another project in which they were involved. Since I had no role in that project, there was no need to invite me to the new meeting. This explanation made perfect sense and I was thankful that I had not reacted in accord with how I had initially felt.

To be honest, it took me a long time to get to the point of being able to take a step back and examine my feelings before acting on them (I suspect that some of my colleagues will say that I still have a ways to go). I also had to learn that although acting in a hostile or withdrawn manner was the most overt reaction to my feelings, just getting those reactions under control was not always sufficient. Reactions to feelings can take more subtle forms as well. Even behaviors such as being less participatory than usual in a meeting can be a reaction to feeling excluded, and making a seemingly harmless but snide comment to a colleague can be a reaction to feelings of hostility. The behavior may be less extreme, but the consequences can be every bit as negative. The more subtle the reaction, the more practice you will need to recognize it for what it is.

The key is to develop an awareness of your feelings and sensitivities. In my case, a tendency to take things personally and a willingness to make assumptions about others' motivations were also contributing factors. Self-awareness is the first step to managing yourself in tough or uncomfortable situations. Chapter 3 shows us that this ability to monitor your internal thermostat is also a key element to becoming an effective negotiator.

Before we get into how you can develop your self-awareness and use it to your advantage, another illustrative case study follows.

Every organization seems to employ someone who is an interpersonal and managerial disaster, but whose knowledge or skills make him or her seemingly indispensable. The situation is often complex: On the one hand, this person may indeed possess unique knowledge and experience, and thus must be kept involved in projects. On the other hand, team members able to make important contributions may be excluded because of the overpowering character of this person. How can an organization retain the input and participation of the domineering manager with low self-awareness while limiting the damage he does to others and to the group? The following case study illustrates one approach.

In this case, a senior scientist behaves in ways that are damaging to his colleagues and company. Because the scientist did not possess good self-awareness, his company had to find a creative way of retaining his expertise while removing him from a position in which he could hinder the work of others.

▸▸ *Case Study: The Clueless Systems Analyst*

Dr. Raju is a founding member at Triotech, a leading designer of cutting-edge enterprise software. He knows the Triotech business inside and out and believes that he knows the best ways to adapt Triotech's products to a company's needs. He has been promoted within the organization over the years so that he now manages a group of midlevel employees. The problem is that morale in his group is low. Raju is domineering in team meetings; he belittles others and provides support only when colleagues agree with him. Alex, a senior systems analyst, feels so put down by Raju that he rarely speaks in team meetings any more. The last time Alex tried to disagree with Raju, Raju told him that he was being "stupid and he needed to go back to school."

The difficulty is that Raju does possess a lot of knowledge and experience, but people will seek his advice only in the most extreme circumstances because they would rather not deal with him. Junior and midlevel staff are dissatisfied with this situation, but management is at a loss about what to do.

Triotech recently hired Elizabeth, a new vice president who was told that she needed to deal with the "Raju problem." Elizabeth spent some time getting to know Raju to understand why he was behaving in this way. Conversations with members of the executive team led her to the hypothesis that Raju had long felt bitter about not being promoted to the position of vice president. He sorely wished to be a member of the executive management team and felt excluded from the inner workings of the company. Management viewed Raju as a "loose cannon" with no executive or strategic capability. Elizabeth suspected that Raju felt underappreciated and his actions in the group environment were a reaction to this. She also thought that Raju might be humiliating others to bolster his own sense of importance, in essence, saying to senior management, "look how knowledgeable I am."

Elizabeth had had some experience with people like Raju and thought that she might be able to help him gain some insight into his behavior and its consequences. However, after several conversations with Raju, Elizabeth concluded that he had little insight into his own motivations and behavior. When opportunities arose for Elizabeth to discuss Raju's behavior and motives with him, he reacted with resistance, deflection, and hostility. When she suggested that Raju examine how he behaved in a meeting from which they had just emerged, Raju exploded in anger and said that his behavior was not the problem; the problem was the stupidity of the people he had to work with every day.

After a number of such attempts, Elizabeth concluded that in the short term, neither Raju's self-awareness nor his behavior was likely to be improved by mentoring or counseling, either by her or someone else. Therefore, Raju could not remain in his current role.

The solution that Elizabeth crafted was to suggest that Raju be promoted to a slightly higher position, something akin to a senior adviser to the executive team. In this new role, he would be invited to many of the senior staff meetings from which he had previously been excluded. However, he would also be expected to continue to participate in line team meetings as a kind of elder statesman, although he would no longer have line management responsibility. Everyone hoped that Raju would continue to provide all the benefits of his knowledge to the company, feel appreciated and acknowledged by senior management, and cease browbeating team members.

This is what in fact happened, but it took the better part of a year for the transition to stabilize, and not without friction. In the end, members of the junior staff were promoted to line management positions and felt empowered and recognized. Eventually, Raju felt sufficiently secure to treat his former subordinates with respect, and the company continued to benefit by having access to his skills.

Two of the keys to this success were Elizabeth's determination that Raju's knowledge was critical to the continued success of Triotech and her recognition that Raju lacked the insight and willingness to examine his own behavior and its consequences. She spent time with Raju to determine how important his knowledge was to the future of the organization. In this case, the company was about to embark on a fund-raising campaign. Although she felt capable of managing the process without Raju, she suspected that this would be a mistake. Few others in the company possessed Raju's ability to address in-depth questions about the company's methodologies.

This story shows how a manager who is attentive not just to the technical facets of her job, but also to individual and interpersonal issues, can arrive at creative solutions to difficult problems. In this case, Elizabeth helped the company retain the knowledge of a valuable scientist and created opportunities for junior staff to contribute and excel in ways that they could not have had previously.

The solution was designed to enable the company to achieve its objectives and satisfy some of Raju's interests at the same time. It involved elements of "out of the box" thinking in that it created a new kind of position for Raju, rather than just marginalize him. The solution was meant neither to punish Raju in a vindictive manner nor ignore or reward his disruptive behavior.

However, not all companies will have a vice president like Elizabeth, who found a creative solution to a tough problem. In other circumstances Raju might have been fired, demoted, or marginalized. In these cases, Raju would have paid dearly for his inability to examine his own behavior and its effects on others, and the company would have lost as well. Most frequently, behavior like Raju's is ignored because the scientific leadership of the company is either reluctant to confront such behavior (see Chapter 4 for more on conflict avoidance in science professionals) or fails to see its destructive nature.

Many of us have known people like Raju, who manage others in destructive ways and are unable to examine their own behavior and motives. If we manage such people and allow their behavior to continue, we own responsibility for the consequences just as much as do the Rajus of the world. Some people who act destructively will be amenable to coaching. Others will need to be moved into positions with little or no line management responsibilities. Others will need to be separated from the organization.

Often, when we see a difficult scientist getting away with the kinds of behaviors that Raju was exhibiting, you can bet that part of the problem is that management does not know how to handle him. In a sense, management is authorizing that scientist's destructive behavior. This typically happens when management feels that it has few alternatives or does not even recognize the presence of the problem. In these cases, the root of the problem may rest as much with management as it does with the scientist. Managers may feel held hostage by someone who possesses domain expertise in areas in which they are ignorant. Management's approach to such problems must begin with an acceptance of its own complicity in the scientist's behavior. If appropriate, coaching and mentoring can be offered. If the scientist is not amenable to these approaches, other tactics, such as the one devised by Elizabeth, can be used.

From our perspective, Raju's difficulty in examining or changing his behavior was due to his poorly developed self-awareness. Greater self-awareness allows us to monitor our feelings and behavior to ensure that they are appropriate to the circumstance. This increases the likelihood of having productive and appropriate interactions with employees, colleagues, and supervisors.

For managers like Elizabeth, the ability to notice or assess a scientist's self-awareness helps them to assign appropriate roles within a team or department. Managers with the ability to help scientists develop their self-awareness are in a position to assist promising scientists develop into leaders (Chapter 9 deals with this theme in more detail).

Scientists such as Raju are not uncommon. From my experience and the research cited in the previous chapter, scientists as a group have low self-awareness. Their inclination is to be analytical and rational, and they may believe that their behavior represents a logical response to external events. Despite this belief, scientists are no less prone than others to experiencing strong or troublesome feelings. If they are oblivious to the impact of these feelings on their thoughts and behavior, they will act in ways that cause them and their organizations considerable trouble.

DEVELOPING AND USING SELF-AWARENESS

Developing and using self-awareness, in the sense that we have defined it at the beginning of this chapter, involves the following three elements:

1. **Knowing** what you are feeling

2. **Anticipating** how this feeling may affect your behavior

3. **Deciding** whether the behavior is appropriate to the situation and will help you

Knowing what you are feeling: To solve a problem, you must be able to describe it

At the beginning of the workshops I run for scientists, I ask participants to pair off and take part in a role-playing negotiation exercise. I deliberately choose topics for negotiation that most scientists dread, for example, negotiating over first authorship for a scientific publication. The groans and eye-rolling from the audience when I announce this topic attest to their discomfort. I describe the hypothetical scenario for the group, making it as difficult as possible for them to come up with an easy solution. I then give them ten minutes to negotiate an agreement.

After the ten minutes, I ask how the negotiation went. Since I have not already discussed the principles of negotiation by this point, the outcome is predictable: Very few reach an agreement.

I also ask what they felt during the negotiation because I know that much of the difficulty people have in tense situations stems from primal behavioral responses to anxiety and feeling threatened. Of course I could see what they feel as I walk around the room while negotiations are taking place: discomfort and anxiety. Responses to my inquiry of how they feel usually consist of the following:

- I wanted to win

- I felt that we were stuck

- I could not get my point across

- I could not think of any good compromises

- I felt that my partner was inflexible

After I patiently write these phrases on a flip chart, I tell participants that these are not descriptions of feelings at all; they are descriptions of thoughts. I typically try very hard to get them to tell me what they felt, not what they thought, and I usually do not have much success.

After one such workshop, I relayed my frustration at not being able to get these scientists to describe what they were feeling to my co-author, Suzanne Cohen, who immediately diagnosed the problem. She said that the scientists could not describe what they felt because they were not accustomed to using the language of feelings, which is somewhat like using a foreign language—if you do not know the vocabulary, you cannot describe what you are experiencing. Moreover, she pointed out something even more profound: If you do not have the vocabulary to describe what you are experiencing, you may not even be aware that you are experiencing it.

What were the words that the workshop participants could not use? They were words such as anxious, scared, threatened, angry, and hurt. These describe powerful feelings and emotions that have profound effects on behavior. If my workshop participants were not aware of their feelings and could not name them or put them into words, what were the chances that they could recognize the impact of their feelings on their behaviors?

The ability to monitor one's own emotions in real time is a key element of self-awareness. Self-awareness is a prerequisite to anticipating your reactions in response to strong feelings, some of which may be inappropriate to the situation. This skill is essential for functioning in tense and emotion-laden situations such as negotiations (discussed in the following chapter) and team meetings.

Now when I conduct workshops for scientists, I list "feeling" and "thinking" words (see below) to help them see the distinctions and get used to using such words.

Thinking words	Feeling words
Want to win	Threatened
Determined	Frustrated
Strategizing	Angry
Not getting point across	Hurt
Stuck	Sad
Not being listened to	Happy
Anything that begins with "I felt that..."	Anxious

A key to recognizing a thinking phrase is anything that starts with "I felt that... ." Even though this sounds like a description of a feeling, "that" is the tip-off that it is not. See for yourself how different it feels to say "I felt that I was in danger" compared with "I felt scared." The first puts you at a distance from the feeling, whereas the second expresses the feeling.

If the ability to notice what you are feeling in real time is the first step to increasing your self-awareness, naming the feeling is the first step to noticing it. The exercises at the end of this chapter will help you to learn how to accomplish this in everyday situations.

The second step to becoming more self-aware is anticipating how you will react after you experience a strong feeling. For example, as illustrated in the first case study in this chapter (Left Out), I know from past experience that my reaction to feeling rejected is to withdraw and pout. I also react by developing imaginary scenarios to get even with the person whom I believe rejected me. In the past, I would actually implement these plans (which, paradoxically, usually involved ignoring those whom I believed had slighted me). Over time, however, I learned that this is how I react to feeling rejected, and I learned to anticipate my reactions as soon as I experience the feeling.

I took control of my behavior because I was able to notice and identify my feelings and know that my initial reactions to these feelings might be inappropriate to the circumstances. Note that even if I had been intentionally excluded from the meeting, my initial reactions would still have been inappropriate. There are many effective ways in which I could have dealt with being intentionally or inadvertently excluded from a meeting to which I should have been invited. If my goal was to solve the problem of why I was not invited, and to make sure that it does not happen again, I could have met with the meeting organizer to find out why my presence was not needed. Chances are that there would have been a good explanation. Becoming hostile or withdrawn would not have helped me to achieve my goal. It is almost always more effective to ignore insults or slights, real or perceived, and remain focused on your mission and objectives (see Chapter 3).

Anticipating how feelings will affect your behavior

Anticipating how you will respond to a feeling requires that you study and learn your own pattern of responses. In the case study about being excluded from a meeting, I anticipated that my feelings of rejection would lead me to withdraw from the group. Others react differently to feelings of insecurity or fear of being marginalized, and the following case study illustrates how someone else might have responded.

▸▸ *Case Study: Zombie Anxiety*

Ted was part of a three-person task force empowered to reorganize the research division of Zombie Biotech in conjunction with a team of external consultants. Ted was anxious that the reorganization would result in his loss of power and influence. This anxiety caused him to control group meetings and meet surreptitiously with the consultants without other team members. Eventually these behaviors came to the attention of the senior vice president of the division, who called Ted to task for his behavior. He told Ted that his behavior was disruptive to the project and promoted mistrust among the other team members. He threatened to remove him from the task force unless he became be more of a "team player."

The paradoxical turn of events in this case was that Ted's behavior in response to his fear of losing influence almost resulted in his removal from the task force. Ted was blindly following his feelings of insecurity by trying to control the situation. Possibly, he had good reason to feel insecure; he may have learned that the senior vice president was unhappy with his performance. It is also possible, however, that Ted's fears were groundless and he may have had a history of feeling insecure in dynamic situations. Whatever the case, Ted's response was clearly counterproductive.

If Ted wants to change his behavior before it affects his career he first needs to recognize and name his feelings, which in this case were insecurity and feeling threatened.

The second step is to anticipate what these feelings might lead him to do. This is not as hard as it may seem; typically, this is the first action that pops into one's head when experiencing those feelings. Ted can use his past experience to recall how similar situations and feelings have caused him to react.

The self-awareness exercises at the end of the chapter can help you to identify your feelings in specific situations and examine what you are likely to do in response, before you do it.

Deciding: Determining the appropriate response

Once Ted recognizes and names his feelings, and anticipates what he might do in response to them, he can decide whether his anticipated behavior is appropriate to the situation. This is perhaps the most important step in the process. Where most of us get into trouble is when we act according to how we feel—often without considering the consequences. Ask yourself the following questions to assess the appropriateness of your reaction to a situation:

- Will my behavior have a negative effect on others, i.e., will it harm, insult, or embarrass?

- Is what I feel like doing inappropriate to the situation?

- Is what I feel like doing inconsistent with my professional responsibilities?

- Am I about to act out of fear or anger? Is my behavior vindictive?

If you answered "yes," even once, think twice about your behavior. One of the guidelines I offer in my negotiation workshops is that if you are imagining how vindicated or self-satisfied you are going to feel after you say or do something, don't do it. Of course, it is necessary to be sufficiently self-aware to recognize the fact that you are anticipating feeling this way.

It takes some experience and practice to evaluate the appropriateness of your behavior. In the example above, even if Ted had taken the time to examine his behavior, he may not have recognized that it was going to have negative consequences. Had he been working on self-awareness for some time, he might have recalled clues from previous situations in which he reacted similarly and which caused negative consequences. Or he might simply use this situation as an opportunity to learn how not to react in the future.

THREE COMMON DICHOTOMOUS REACTIONS TO DIFFICULT SITUATIONS

Just as having the language to name feelings helps you to become aware of them, so too does the ability to name common behavioral patterns help you to become aware of them. The following three questions can assist you in recognizing and naming some common reactions to difficult situations. The reactions are listed in pairs that are behaviorally opposite from one another.

1. **Retreat/attack.** When threatened, do you tend to retreat or attack? Many people behave passively when threatened: They become silent or withdrawn or attempt to leave or avoid the situation altogether. Other passive responses might include ignor-

ing or not speaking to someone who hurt or insulted you. Others become hostile, angry, and threatening. Observe yourself in uncomfortable circumstances and note whether your behavior fits either of these categories.

2. **Verbal/nonverbal.** When you find yourself in an uncomfortable situation do you react verbally or nonverbally? Some people launch verbal defenses or attacks whereas others express their discomfort or hostility nonverbally. Avoiding eye contact, making facial expressions that show your feelings, or moving or holding your body in either aggressive or submissive postures are all nonverbal modes of behavior. Observe whether you manifest these behaviors and use that information to get a better reading of your feelings as well as how you come across to others.

3. **Internalize/externalize.** Do you have a tendency to blame yourself (internalization) or others (externalization) for problems? When some people arc criticized, or when they encounter a problem, they automatically blame themselves. They are ready to take responsibility and feel guilty for everything, regardless of whether it is actually their fault. In contrast, some may shift blame and responsibility to others, even in cases in which most or all of the responsibility is theirs. Observe your own behavior to learn if you have a tendency to behave in either of these extreme ways. This can help you to anticipate what might be an inappropriate reaction the next time you find yourself in such a situation.

The final exercise at the end of this chapter makes use of these behavioral polar opposites to help you identify and remember your own behavior patterns.

HELPING OTHERS TO IMPROVE THEIR SELF-AWARENESS

Thankfully, not every scientist is like Raju. Many are receptive to feedback and with coaching can use that feedback to change their behavior and improve their self-awareness. In the following case study, a manager uses feedback to help a capable scientist become a more effective team member and eventually a team leader.

▸▸ *Case Study: The Angry Scientist*

Several years ago I was advising Colin, a senior scientist, who frequently lost his temper and lashed out at colleagues during meetings. He just could not seem to control himself. Frequently, during discussions of scientific results at team meetings I could see him begin to squirm in his seat. He would become more and more agitated until he could take it no longer, and he would release a scathing critique of whatever was being presented. His behavior typically resulted in hostile responses from others, and the meetings would often degenerate into shouting matches.

This behavior was, of course, hurting Colin's career. Discussions about appointing leaders for new project teams often included Colin, but he would never be chosen. He had made too many enemies and no one was confident that he could lead a team in a constructive manner.

After an especially confrontational outburst, I privately asked Colin why during the meeting he had shouted that a colleague's work was sloppy and useless. His response was, "Shoddy work makes me furious! I can't stand it and I can't lie about it! I have to be honest!" Colin knew that his outbursts were hurting him professionally, but he refused to compromise his scientific and personal integrity by keeping his opinions bottled up. He believed that he owed it to himself and the company to be honest.

I explained to Colin that as a senior scientist he was expected to use his knowledge to solve problems, and that shouting at colleagues was not the way to do this. I asked Colin whether he could express his scientific concerns in a helpful manner, as opposed to a hostile one. He was astonished by my question. He said that by expressing his hostility he was behaving honestly and could not see anything wrong with that. He felt that not expressing what he felt, or keeping his fury bottled up, would be duplicitous.

After he said this, I saw a look of confusion settle on his face. I suspected that he was making the connection between his insistence on being honest about his feelings and the disastrous outcomes for him personally and for the group as well.

I asked Colin why he thought that it was his responsibility to express his every feeling. I suggested to him that the issue was not honesty but rather one of deciding whether to act on every feeling that he happened to have. I reminded him that it was one thing to feel angry and quite another to act angry; the decision was his. It had nothing to do with honesty or duplicity and everything to do with choosing a behavior that was most likely to get the job done.

Somehow this line of reasoning had never occurred to Colin, yet I could see that it made a lot of sense to him. I also told him to think about what he hoped to accomplish by criticizing the presentation in his aggressive way. I reminded him again that his role in the company was that of senior scientist, not senior arguer. His responsibility was to ensure that good science was done. To accomplish this, he needed to provide his expertise to his colleagues in a way that they would hear and accept. Behaving in a hostile, argumentative manner would not accomplish this.

Colin thought it over and agreed that choosing his behavior, rather than simply acting on his feelings, allowed more control over himself and the situation. I suggested that the next time he felt himself getting furious with someone's presentation he just sit there, remain quite, and pay attention to what he was feeling. Once he experienced his anger consciously, he could then have an internal dialogue with himself about the consequences of expressing that anger. I also suggested that at his next meeting, he try to make just one comment that reflected his scientific concerns in an emphatic but emotionally neutral manner.

Once Colin took control of his feelings, his behavior improved—so much so that he was eventually promoted to group leader. Over time, Colin became far more aware of both his own feelings and their effect on others, which in turn helped him to control both. Colin was one of the most rigorous scientists I have ever known, yet he was also very open to feedback about his behavior and to discussing it and his feelings.

Had Colin not been so open, my suggestions might have fallen on deaf ears or been met with hostility and denial, like Elizabeth's discussions with Raju. There is no foolproof way to predict how people will react when presented with an opportunity to examine themselves. But if you ease into the discussion slowly, you are likely to sense where the barriers and defenses in different people's personalities lie. Honing your powers of observation by paying attention to what others say and their body language (see Chapter 5) is a good place to start. Chapter 9 presents a list of questions to ask yourself about others that may also be useful in gauging someone's capacity for self-examination.

Finally, do not try to help others become more self-aware unless you are willing to examine your own behavior and motivations as well. You will appear hypocritical if you cannot be as open about your own behavior, and as receptive to observations about yourself, as you wish others to be. Moreover, the other person's behavior has usually not arisen in a vacuum. Behavior is always influenced by or is in response to the behavior of others, and in some cases of yourself. Peter Senge, in his book *The Fifth Discipline* (Senge 1990), notes that "Trying to fix another person's defensive routine is almost guaranteed to backfire... . By focusing attention on the other person, the "confronter" has taken no responsibility for the situation... . If we perceive a defensive routine operating, it is a good bet that we are part of it."

Before you try to help others with what you perceive to be their problems, be sure that you have given careful thought to your own role in creating, facilitating, or exacerbating those same problems. My complicity in the preceding case study might have involved giving Colin subtle messages that I also was annoyed by shoddy data and that I secretly relished his acting as my proxy in browbeating other scientists. I might also have discovered that I had a tendency to assign projects that were clearly beyond the scientist's capabilities. In that case, Colin's annoyance and frustration might have been directed against the unwitting scientists because he could not express his anger to me directly. In either case, my ability to reflect on and recognize my own complicity in Colin's behavior was a prerequisite to my conversation with him.

USING SELF-AWARENESS TO HELP GUIDE BEHAVIOR

Like the Raju case study, the one about Colin had a positive outcome, but for very different reasons. Colin had enough self-awareness to recognize his feelings and enough empathy to understand the effects of his behavior on others. Building on this understanding, he was able to change his behavior dramatically, and he was rapidly advanced in the company. Raju, on the other hand, had no such insight. Although he had to be removed from operational responsibilities, the company found a role for him in which it could access his expertise and his behavior would do minimal damage. The point is not that one outcome is inherently better than the other, but that both were achieved by understanding the impact of self-awareness on the scientist's ability to change and adapt. Without this perspective, both Raju and Colin might well have been fired or marginalized, to their own and the company's detriment.

By using self-awareness, you can gain greater control over your actions and better align what you do with the needs of the situation. The facility to become aware of your feelings and their behavioral consequences requires practice, as does the ability to decide whether the anticipated behavior is appropriate or dysfunctional. The following exercises are designed to help you acquire these skills as well as practice and improve them.

REFERENCES

Goleman D. 1995. *Emotional Intelligence*. Bantam Books, New York.

Goleman D., Boyatzis R., and McKee A. 2002. *Primal Leadership: Realizing the Power of Emotional Intelligence*. Harvard Business School Press, Boston, Massachusetts.

Senge P.M. 1990. *The Fifth Discipline: The Art and Practice of the Learning Organization*, p. 255. Doubleday/Currency, New York.

EXERCISES AND EXPERIMENTS

These exercises are designed to help you to develop self-awareness in three areas: thoughts, feelings, and sensations. You will become aware by paying attention and noticing, and then registering what you notice. There are no right or wrong answers or behaviors. The objective is to be descriptive, not judgmental, about what you observe. The exercises are best done in a quiet place where you will not be distracted, but see the instructions below for additional location options. Pay full attention to very specific parts of yourself.

1 Notice what you are thinking

Thoughts can take the form of words, images, or memories. They can also be about what has already happened or what will happen, and important or trivial matters.

Start by recording your thoughts after you notice them. You may find it interesting to do this under several different sets of circumstances. First, try it alone in a quiet place. You can also try it in the presence of others—perhaps during a meeting in which you can periodically take a moment to listen to and take note of your thoughts. Finally, try it during a tense or threatening situation—perhaps a difficult discussion in a meeting or a tense phone conversation. In these circumstances, you may be able to tune into yourself for brief moments without interrupting your overt participation.

Needless to say, it is probably not a great idea to take these momentary "time-outs" during a critical negotiation about your salary or a promotion. You will eventually get to the point at which you can do that with practice, but for starters try it during situations where being momentarily disengaged will not be detrimental to you or the meeting.

What is the content of your thoughts? Are they about things, people, or events? Are they about planning for the future? Going back to past events?

2 Notice what you are feeling

Naming feelings gives us more control over them. This exercise consists of writing down what you are feeling as you notice them. You may wish to record your feelings in the following format: Right now I am feeling *x*. Yesterday I felt *x* when *y* happened. Today I feel *x* when *y* happened. For example, "Today I felt relieved when I my experiment worked."

Again, practice this exercise in a variety of situations, especially ones in which you anticipate feeling uncomfortable. Ease into the process by taking momentary time-outs. At first, you may find the process distracting, but as you become more adept at identifying your feelings, you will need increasingly less time to access them. To help you find the right words to best describe your feelings, we have assembled a word list below. Of course, the list is not all-inclusive, and you may wish to add others.

Anger

aggravated	enraged	judgmental
aggressive	envious	mean
agitated	exasperated	miffed
angry	frustrated	miserable
annoyed	furious	negative
argumentative	grouchy	oppositional
burned up	grumpy	pissed off
confrontational	hostile	resentful
contemptuous	impatient	spiteful
disapproving	irritable	stubborn
disgusted	irritated	suspicious
dislike	jealous	teed off

Fear

agitated	fearful	nervous	tense
anxious	frightened	pressured	threatened
apprehensive	horrified	pushed	undecided
blocked	in doubt	reluctant	upset
cautious	irritable	restless	vulnerable
dissociated	jittery	scared	worried
dreading	jumpy	shy	
edgy	mistrustful	spaced out	

Happiness (and other positive feelings)

affectionate	confident	hope	relieved
amazed	contented	hopeful	respected
appreciated	delighted	in awe	satisfied
aroused	determined	interested	sensual
assertive	eager	joyful	sentimental
bold	elated	kind	soothed
brave	empathic	masterful	sure
calm	enjoying	optimistic	sympathetic
capable	enthusiastic	passionate	tender
cared about	euphoric	peaceful	thrilled
cared for	excited	pleasant	triumphant
close	exhilarated	pleased	warm
compassionate	glad	powerful	
competent	happy	proud	

Sadness (and other negative feelings)

alienated	burdened	envious	lazy
alone	childish	foolish	let down
apathetic	confused	humiliated	lonely
ashamed	defeated	hurt	lost
bad	defensive	indecisive	moody
bashful	depressed	insecure	restless
betrayed	disappointed	insulted	sad
blocked	discouraged	isolated	weak
bored	disturbed	jealous	

3 Notice your sensations

Sometimes we can gather important "physical" information that is otherwise unavailable to us through thoughts or feelings. This exercise consists of recording sensations as you notice them.

Sensations are in the present, not the past or future. Sensations include warm and cool, vibrations and tingling, discomfort and comfort, lack of sensation and numbness, tension and relaxation, calm and excitation, pressure and lightness, contraction and expansion, hard and soft, increasing and decreasing energy, wet and dry, stiff and pliable, and full and empty.

Sensory experience occurs through the skin, the five senses, and the internal organs. For example, we might notice that our stomach is churning, our heart is beating quickly, our palms are sweaty, or our facial muscles feel soft and relaxed.

Like the exercises for thoughts and feelings, try this one in a variety of circumstances during your work day. Pay attention to sensations that occur in your body and write them down when you have the opportunity.

4 Putting it all together

The thoughts, feelings, and sensations you experienced in the above situations represent your self-awareness data. With practice, it will become increasingly easier to gather this data as you learn to notice yourself and your experiences. Think of self-awareness data as the raw material required for making better decisions.

Reflect on a situation that you have experienced recently, yesterday or today. Try to reconstruct your thoughts, feelings, and sensations from that experience. Take the time to review your state of mind.

Review the thoughts, feelings, and sensations that you recorded during a situation in which you were taking time out to write them down. Do these observations give you a useful perspective on the way in which you behaved (or wanted to behave) or what you said (or wanted to say)? Do these thoughts and feelings help you to make sense of how you acted and what you said in similar situations in the past?

If you have an interaction that causes you to get "off balance," recall your sensations, feelings, and thoughts by bringing your attention to them. This requires only seconds as you check in with yourself. You will gather important information that will help you to regain control.

5 Know your hot buttons

"Hot buttons" are sensitivities that cause you to experience strong feelings in response to something that someone says or does. A key element of self-awareness is knowing in advance what causes you to feel upset, angry, hurt, etc. The importance of this knowledge cannot be overestimated, especially if you are prone to respond in a manner that is not in your best interest.

To recognize your hot buttons,

- Recall past incidents in which someone has done or said something that elicited an extreme reaction or a strong feeling. Write down what specifically was said or done to trigger that reaction. In your own words, or from the list above, record the feelings you experienced that were associated with your reaction.

- Categorize the type of stimulus that triggered the reaction. Were you accused of lying or incompetence or were you left out of some activity, ignored, or challenged?

- Identify the trigger situations and list as many as possible that have elicited emotional reactions in the past.

- Remember the results of this exercise the next time that one of these situations arises. When you feel a hot button response coming on, do not act, but instead find some way of extricating yourself from the situation, at least momentarily. For example, excuse yourself to go to the rest room, answer an imaginary cell-phone call, or just remain quiet for a moment to give yourself time to notice what your feeling.

- From this new vantage point, ask yourself if reacting to your feelings is the best course of action, if your reaction is appropriate to the circumstance, if you will be making a tense situation worse, and if it is best to cool off before you do or say anything. The more extreme your feeling, the more you must avoid acting on it.

Over time, you will learn to recognize hot button responses as they are happening, or even to anticipate them, giving you greater control over your reaction.

6 Identifying behavioral polarities

The following table presents questions that are designed to help you think about and remember how you are likely to behave in certain uncomfortable or stressful situations. The responses are framed in the form of behavioral opposites. Use your responses in conjunction with the self-awareness exercises on page 30 to help you identify and remember how you are likely to behave in similar circumstances.

	Attack (counter-accusations, trade insults)	_Retreat_ (refuse to take the offensive or to defend yourself)	_Verbal_ (e.g., argue, discuss, inquire)	_Nonverbal_ (e.g., withdraw or express feelings with facial expression or hostile emotions)	_Internalize_ (blame one's self)	_Externalize_ (blame others)
In a threatening situation, my first response is						
In a disagreement with a peer, my first response is						
In a disagreement with my boss, my first response is						
When I feel insulted, my first response is						
When I feel hurt, my first response is						

CHAPTER 3

Gordian Knots: Solve the Toughest Problems through Negotiation

Learning negotiation, persuasion, and diplomatic skills is important for a scientist... . Obtaining these skills is a critical part of a scientist's training and is generally acquired by watching the behavior of others. *Diplomacy is essential for preserving relationships that may be important for a fellow's career development, and a key step is learning how to cooperate with the very people whose help will be needed to achieve goals.*

A GUIDE TO TRAINING AND MENTORING IN THE INTRAMURAL RESEARCH PROGRAM AT NATIONAL INSTITUTES OF HEALTH*

When I run workshops for scientists, my favorite exercise is to have them pair off and take part in a mock negotiation involving first authorship for a publication. I tell them to come up with a solution agreeable to both sides, and they look at me as if I am a lunatic. After all, they say, there can only be one winner. Yet 30 minutes later, after I have shown them how to "expand the pie" during a negotiation, they have become true believers.

WHY LEARN NEGOTIATION?

Scientists who notice and pay attention to their experiences as well as the behavior and reactions of others are better equipped to function and succeed in the complex social world of scientific research.

A surprising number of work-related interactions that scientists have daily are negotiations in one form or another. In fact, we use the term negotiation to refer to almost any interaction in which there is a difference of opinion, the interests of the

▸ **Why learn negotiation?**
What you learn from negotiation

▸ **Why learn principled negotiation in particular?**
Principled negotiation

▸ **Not everything needs to be negotiated**

▸ **Negotiate in good faith**

▸ **Do not enter into a negotiation unless you are prepared**

▸ **The elements of principled negotiation**
Preparing for a negotiation
Always negotiate with "interests," not "positions," in mind
Maintain a collegial interaction by using "I" statements
Learn to expand the pie
Manage yourself: Know your hot buttons and watch your body language
Do not take anything personally during a negotiation
Listen to and acknowledge the other side: Be empathetic
Defuse anger and hostility

▸ **References**

▸ **Exercises and experiments**
1. The right tools for each negotiation
2. Identifying underlying interests
3. Identifying negative listening patterns in yourself
4. Correcting negative listening patterns

*http://www1.od.nih.gov/oir/sourcebook/ethic-conduct/mentor-guide.htm#Interactions

33

participants differ, or the two parties have conflicting agendas. Discussing the design of an experiment, interpreting experimental data, planning for the next experiment, getting access to reagents and equipment, working out the specific contributions of those who are involved in the project, and, of course, every scientist's nightmare, deciding who will be an author of the resulting scientific paper or report, and in what order, are all negotiations.

If you approach these and other negotiations as power struggles or opportunities for manipulation, you may get what you want in the short term, but in the process may create such ill will and resentment that future negotiations will be far more difficult. Alternatively, if you have such anxiety about negotiating that you either avoid it or accede to others' demands at the first sign of conflict, you may find yourself taken advantage of, time after time. In both cases, the consequences of how or whether you negotiate will limit your effectiveness and productivity. Learning to be effective at negotiation can be as important for your success in science as learning to be skilled at organic chemistry, string theory, or making knockout constructs.

As you learn to be a good negotiator, you will become adept at many of the core skills that we present in this book including

- Noticing what you are thinking and feeling
- Anticipating the behavioral consequences of what you are feeling
- Recognizing your hot buttons
- Managing anger
- Identifying underlying interests
- Attacking the problem, not the person

These skills are important elements of successful management in science and technology and it is no coincidence that these same skills are the hallmark of good negotiation. Effective negotiators need to have sufficient self-awareness to choose appropriate behaviors in tense situations. They must be able to avoid responding to inflammatory comments in kind and to remain focused on the issues of the negotiation. They must learn to promote alliances, even when the other person is resistant. Good negotiators also need to accurately hear the concerns and interests of the other person to understand their viewpoint. Finally, good negotiators must observe the effects of their own behavior on others to ensure that what they are saying is heard and understood as intended. Learning to be a good negotiator is a great way to integrate and use many of the most important skills in this book. Over time, as you practice your negotiation skills and internalize them, you will find yourself becoming much more "tuned in" to yourself and others during all types of interactions, and you will manage yourself and others more effectively.

What you learn from negotiation

- Becoming a good negotiator forces you to "read" needs, interests, and beliefs of others.
- Good negotiators learn to monitor and modulate their own behavior in tense and emotion-laden situations.

- Negotiation teaches that listening can be more productive than talking.

- Effective negotiators (like good managers) identify and focus on underlying interests rather than on rigid positions.

WHY LEARN PRINCIPLED NEGOTIATION IN PARTICULAR?

Of the many approaches to negotiation, the one we advocate most is often referred to as "principled negotiation." One of the chief architects and supporters of this form of negotiation is William Ury, who with Roger Fischer wrote the book *Getting to Yes* (Fisher and Ury 1991) and subsequently, *Getting Past No* (Ury 1993). Principled negotiation focuses on the interests and needs of yourself and the person or people with whom you are negotiating.

The goal of principled negotiation is to arrive at an outcome that satisfies as many of the interests of both parties as possible.

We should note that there are many other approaches to negotiation and it is not uncommon to find books on negotiation in airport book racks, with such titles as "Learn to get everything you want through negotiation" or "How to be a killer negotiator and give up nothing," etc. My advice is to avoid these approaches like the plague. As a scientist, most or all of your negotiations are with professional colleagues, employees, employers, and others with whom you interact on an ongoing basis and have long-term relationships. Whether you know it now or not, these relationships are important to you and your future. The last thing you want is to use a bag of clever negotiation tricks to convince these people to come to an agreement that they may regret or that leaves them with the feeling of having been manipulated. In fact, in general, most of us are not very adept at being clever during a negotiation. It takes too much effort, and as a result, we almost always come across as devious and insincere. Most people can read these signs from a mile away and react defensively, with suspicion and caution. This is a nonproductive way to interact with professional colleagues and should thus be avoided. What you learn here will be less about tactics and strategy—important aspects of some types of negotiations—and more about reaching an agreement while maintaining or strengthening the relationship with the other person.

In the following, we present those elements of principled negotiation that are most relevant to the themes of this book and to scientific or technical settings. We recommend that you read *Getting Past No* (Ury 1993) for those negotiating elements not covered here, such as crafting an agreement that will last, helping the other person find a solution, and demonstrating the consequences of not reaching an agreement in a nonthreatening way.

Principled negotiation

- Participants work to satisfy the underlying interests (not the positions) of each.

- Negotiators focus on interests, which leads to creative solutions.

- Participants are forced to create new options ("expand the pie").

- Assumes a mutual expectation for an ongoing relationship between negotiators.

Underlying assumptions

- Conflict is a normal part of human interaction and need not be avoided.

- Conflict can be resolved in a collegial manner.

- Participants are problem solvers, not adversaries.

- Interests of all parties deserve respect and consideration.

- Participants avoid rigid positions.

NOT EVERYTHING NEEDS TO BE NEGOTIATED

Before I discuss the fundamentals of principled negotiation, let us address one question that scientists ask during every workshop I run. I am invariably presented with some situation that calls for a difficult decision on the part of the presenter, who asks, "How do I negotiate that?" The situation may have to do with a scientist not performing her duties or an important decision that needs to be made quickly. My response is that not every situation calls for negotiation. You as a leader just have to use your judgment to decide on these situations. They often include instances in which you have the undisputed authority to make a decision or the decision needs to be made quickly or by you alone, because involvement of the other party would be inappropriate. An example of the latter might be excluding a troublesome employee from a discussion involving the type of disciplinary action to be taken against him.

Another circumstance in which you must question whether negotiation is appropriate is when you are being subjected to disrespectful or abusive behavior. If you cannot put a stop to abuse by being direct ("I'm feeling very insulted by your accusations and we will need to continue this discussion at another time unless it stops"), you always have the option of walking away.

Be cautious about starting a negotiation with someone who has no intention of being fair or considerate of your interests. In addition, be careful if the other party is in a position of strength at the outset of the negotiation and you are having a hard time coming up with convincing arguments to support your views. Situations such as these present a good case for putting off the negotiation until you are better prepared, or until the other party shows a willingness to address your interests as well as their own. If you choose your negotiations carefully, as well as the time and circumstances under which you engage in them, you will have a greater likelihood of satisfying your interests.

NEGOTIATE IN GOOD FAITH

Sometimes you will have the authority to make a unilateral decision, but you will choose to negotiate as a way of ensuring the buy-in and involvement of the other party in the outcome. If you enter into such a negotiation, do so in good faith and with an open mind.

There may be cases in which you sincerely believe that the decision could be negotiated but the outcome is so important that you need to make it unilaterally. Scientists more than others may be more prone to such an attitude, especially if they believe that because their views are based on solid science, there is nothing to discuss. If your intention is to make a unilateral decision, make this clear from the outset. Many people confuse trying to convince someone of their point of view with negotiating. For example, you may believe that you know how to correctly conduct an experiment, but your postdoc has a different view. You invite the postdoc into your office to "discuss" the matter, but your intention is actually to convince him to do it your way. Your plan is to appear open, but in your mind, the outcome has already been decided.

This is one of the most common errors that people in positions of authority make during a negotiation. If you impose a predetermined unilateral decision because you cannot convince the other side of your position, you have changed the interaction from negotiation to manipulation. Unless she is naïve or ignorant, the other person will almost always see what you have done. The consequence will be a lack of trust during future negotiations.

If your mind is made up and you have no interest in hearing more on the matter, say so in a polite but clear manner, but do not dissemble. You might say, "Janice, I've decided that the experiment needs to be done on rats, not rabbits. I know that you have a different view and I respect that. But in this case, I'm going to make this decision. We can see how it goes, and if I'm wrong we can decide how to proceed from there."

DO NOT ENTER INTO A NEGOTIATION UNLESS YOU ARE PREPARED

Many times you will find yourself knee-deep in a negotiation that you had no intention of having. We often get surprised by a discussion that quickly turns into a contentious struggle. In this situation, first become aware of what is happening. If you find yourself responding instinctually to a threatening discussion, try the techniques described in Chapter 2 to gain some psychological distance on the situation. Also, use the tools in Chapter 5 to calm the other person down if they are angry or confrontational. Once you have done this, you are in a position to make a decision about whether it is in your best interest to continue with the impromptu negotiation or to postpone it until you have had time to gather your thoughts. You will most always have a better outcome if you can do the latter. If you need or wish to continue with the negotiation, the more familiar you are with the following guidelines, the more confident you will be that the other person will not take advantage.

THE ELEMENTS OF PRINCIPLED NEGOTIATION

Preparing for a negotiation

Before going into a negotiation, become thoroughly familiar with all of the facts and issues, even if negotiating with a good friend or colleague. Good friends can make bad agreements, if only because one of them forgets to bring up some important issue that should have been included in the discussion. What seemed like a fair and equitable agreement at the time ends up feeling like a poor agreement later. Then you are faced with either a bad agreement that you have to live with and may resent or the prospect of reopening the negotiation later.

In *Getting Past No* (Ury 1993), Ury outlines the basic elements of preparing for a negotiation. These include the following five components.

i. Identify your and the other person's interests underlying the negotiation

Interests are the underlying needs and wants that you hope will be satisfied by the negotiation. As we will discuss in more depth later in this chapter, interests differ from positions. Most people go into a negotiation with a position, e.g., "I need to use these departmental funds to expand my tissue culture facility." In this example, the interests may include ensuring that a particular set of planned experiments can get done, increasing the throughput of your lab, or replacing an older facility that is difficult to maintain. Before you enter into a negotiation, determine the interests that you will try to satisfy. As much as possible, do the same for the other side. This will help you to be more creative when coming up with ways to satisfy those interests during the negotiation.

ii. Identify multiple ways in which interests can be satisfied

Once you have identified both parties' interests, you are in a good position to consider the options for satisfying them. The more options you have, the greater your flexibility during the negotiation. Having a list of options in advance gives you a valuable advantage for proposing effective solutions.

iii. Identify standards that can be used to evaluate proposals

Try to find guidelines, precedents, or paradigms to help determine the fairness of a decision or whether it is consistent with similar decisions made before. In some cases, standards can be very useful and even decisive in a negotiation. For example, standards about who gets included in authorship of a scientific paper can help resolve such thorny issues. Institutional salary guidelines or salaries from other institutions can be valuable information to have during salary negotiations. Do not overuse standards, though, and do not rely on them as your sole justification for a viewpoint. Unless the standards are universally accepted and unequivocal, you may open yourself to comments such as "just because they did it that way does not mean that we should."

iv. Think through your alternatives to a successful negotiation

No matter how well prepared and how flexible you are, it may not be possible for you to reach an agreement in a negotiation. In such a case, you have the option of walking away without an agreement. If you find yourself in such a situation, have a good understanding of the consequences. In scientific settings, which are often in the context of large research organizations, the most common fallback scenario will be that some higher authority will intervene and make the decision. If this person is perceived by both parties as neutral, this may be an acceptable solution. In some cases, however, you may have no acceptable alternative. Using the above example, if you and your supervisor are disagreeing over whether rodents or rabbits are the better experimental model, you may have no choice but to go along with her decision. On the other hand, if you are negotiating over salary or a promotion, your alternative to an agreement might be to seek a position elsewhere.

v. Define your aspirations and limits

Categorize the kinds of agreements that might arise from the negotiation. Can you find an ideal outcome in which your interests can be met? This outcome may be unrealistic, but it is helpful to speculate what it would be. Next, make a similar list of outcomes that would be acceptable, for example, an outcome that satisfies most but not all of your important interests. Finally, could your interests be satisfied in ways that might be less than optimal, but acceptable? This is your bottom line. Agreements that fail to achieve this level of satisfaction will be unacceptable, and trigger you to walk away from the negotiation and exercise whatever alternatives are available. If possible, try to perform the same exercise for the other person by putting yourself in his position. Having thought through what they might view as acceptable or unacceptable will help you to craft an agreement that is mutually satisfactory. If you feel that you do not have sufficient insight into the other person, you may not be paying sufficient attention to what he says and does during your interaction. See Chapter 6 for suggestions on improving this.

Always negotiate with "interests," not "positions," in mind

An interest is what you ultimately try to achieve in a negotiation. Positions are specific ways in which interests can be satisfied. This is the single most important lesson in this chapter. If you learn nothing else, learn this. Most people negotiate to get something that they have decided they need or want: a salary raise, a mass spectrometer, someone's approval of a project plan, etc. These people typically enter into a discussion with a specific objective or outcome in mind, such as "I deserve a $10,000 raise," "I need you to approve this project plan I wrote," or "We need to buy this $250,000 mass spec for the proteomics group." In each case, the objective is expressed as a position. They have decided what they need, and they seek to convince the other person to go along with them. This approach feels especially natural to science and technical professionals because they typically start by analyzing a problem and then come up with a solution. All they need is to convince the other person to buy into their solution.

The problem with this approach is that staking out a position limits the other person's spectrum of responses. If the other person has a problem with your position, you have actually made it easy for them to say, "We don't have the money for a raise," "I cannot approve the plan as it stands," or "Use someone else's mass spec." Suddenly you are arguing for your position and the other person is resisting.

Never go into a negotiation with positions. Instead, go in with interests in mind. To discover your interests, you need the capacity for self-reflection that we discussed in the previous chapter. Going into a negotiation with demands or rigid positions means that you have not examined your motives and the thoughts and feelings that influence them. Only by stepping back and uncovering the underlying interests that you believe your negotiating position will satisfy can you escape the rigidity of positional thinking. Recall the lessons of Chapter 2 in which you learned to examine and identify your feelings. Knowing your feelings can be a great way to identify your interests. In many cases, you may think that your negotiating position is the result of a rational analysis of all possible alternatives, when in fact it is actually the result of choosing an outcome that "feels right" without actually understanding why.

Let us examine in detail three examples of positions and their underlying interests, and ways to satisfy those interests that go beyond the initial position.

Problem 1

Positional thinking. "I need to purchase an additional mass spec for the proteomics lab to keep pace with our work requests. I know that it's not in the budget, but we're swamped."

Analysis. Why do you think that you need to buy a new mass spec? Perhaps people are lined up at your door with samples to be analyzed and you continually put them off, explaining that your group is backlogged. You are under a lot of pressure. You have analyzed the workflow in the facility and can find no way to increase throughput. You decide that the only way to solve your problem is to buy another mass spectrometer, which costs $250,000. This position feels right because it is a quick fix, and it shows your team that you can go to bat for them and come up with a solution.

What if you had analyzed your underlying interests?

Underlying interest 1. You must meet the demands of proteomics facility users for protein analysis. The following alternatives will satisfy interest 1:

• Send some samples to an outside lab

• Increase hours and/or staff in proteomics facility

• Use alternative techniques to get needed data

Underlying interest 2. Your staff is campaigning for more equipment. They feel that you have not supported them in the past, and you want to show support for them now. The following alternatives will satisfy interest 2:

• The vice present tells your staff that you are campaigning for their interests and you are both working to come up with a solution.

• You find an affordable alternative to show support for your staff, such as providing free pizza and extra days off for anyone who works late helping to meet the backlog.

Problem 2

Positional demand on you. Your vice president for research and development (R&D) says, "We need to have the toxicology studies for this cardiology compound completed in half the normal time. Drop everything else and get this done."

Analysis. The company has suffered a serious setback in one of its major programs. Management is anxious and they want to put their best foot forward at the next board meeting. The vice president of R&D has just spoken with the chief executive officer, who told him to push the most promising programs so that he will have something concrete to show at the meeting. The first thing that occurs to the vice president is the new cardiology compound.

Underlying interest. The vice president is under pressure to come up with data on a product candidate by the next board meeting. The following alternatives may help her to satisfy her underlying interests:

- Offer strong interim, but not final, results

- Suggest that the vice president present nearly complete data for another drug candidate instead, in the metabolic disease arena

- Suggest outsourcing some of the studies to complete them on time

Problem 3

Your positional demand to division chief. "I'd like to talk to you about my salary. I think I did a great job making those knockout mice for the project, but I think that I'm being underpaid and would like to request a $10,000 per year raise."

Analysis. You have worked hard during the past 18 months on an important project to create several knockout mice. The work was successful, but you feel that your boss has used your data without any real appreciation for what you did. You recently heard that a person with a position identical to yours in another division of the company has a salary that is considerably higher than yours. You go to your boss demanding a salary increase, but is this really what you want? This position feels right because you feel unappreciated and getting this raise will be a validation of your importance.

Underlying interest. You feel unappreciated by your boss and would like some recognition. The following alternatives will satisfy your interest:

- Supervisor gets the message and suggests a promotion to a higher level of responsibility with the promise of a salary review within six months.

- Supervisor offers to help you to increase your visibility by having you present the results of your studies at a major company review and an upcoming international meeting.

- Supervisor agrees to compare the company's salary structure with others in the industry and address inequities.

- Supervisor convinces you that there are no funds for increased salaries. He offers to assign another postdoc to work under your supervision. This makes you that feel he appreciates your work and it increases your productivity.

The examples show that in each of these cases, you will have more success if you focus on satisfying underlying interests than on predetermined or rigid positions. The examples show that what seem like simple positions are actually proxies for complex underlying interests, and that each interest can be satisfied in many ways.

The underlying complexity of these seemingly simple positions makes it necessary to think carefully about how you phrase requests for a negotiation. The right words will help you to begin a discussion focused on the important underlying interests. For example, in the salary case, you may say, "I believe that I may not be getting compensated adequately. I'd like to discuss how we can remedy this." Later in the conversation you can add, "If there's no money in our budget for raises, maybe we can look at other options, such as some extra vacation time or the possibility of a promotion to a higher pay grade. What do you think?"

In the mass spec case, you might say, "My team has been backlogged for the past year. We came up with some ways to remedy the situation that I'd like to discuss with you. One of the solutions is to purchase another mass spec, but you may have other ideas that would help."

In the toxicology case, you could try, "Tim, you know that we're under pressure because of the failure of the CD 345 compound last month. I need some positive news for the board meeting next quarter and one idea is to accelerate the cardiology project. Do you have any other suggestions?"

In all three examples, the idea is to get the other party involved in helping you to find solutions that satisfy your underlying interests. Then the negotiation becomes a problem-solving session and not a demand for a specific outcome. Most people respond much more positively to a request for help than to a demand that they buy into your solution.

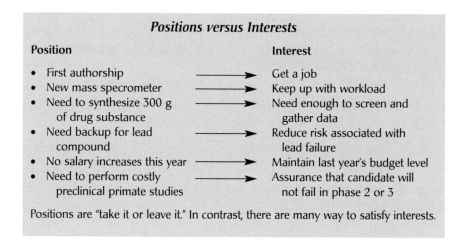

Positions *versus* Interests

Position	Interest
• First authorship	Get a job
• New mass spectrometer	Keep up with workload
• Need to synthesize 300 g of drug substance	Need enough to screen and gather data
• Need backup for lead compound	Reduce risk associated with lead failure
• No salary increases this year	Maintain last year's budget level
• Need to perform costly preclinical primate studies	Assurance that candidate will not fail in phase 2 or 3

Positions are "take it or leave it." In contrast, there are many way to satisfy interests.

Maintain a collegial interaction by using "I" statements

In a negotiation, or in any professional discussion, if you find yourself ready to start a sentence with the word "you," stop and think. "You left my name off this manuscript" may in fact be true, but it introduces the discussion with an accusation. The other party is placed in a defensive posture and may be tempted to respond in kind: "You really didn't contribute very much to the work." The discussion goes downhill from there.

Do not frame the problem in terms of what the other person did, but rather in terms of the impact on the project, or its effect on you. "I see that my name does not appear on this manuscript and I think it should" makes the same point, with the same frankness, but addresses the issue rather than the integrity of the person.

The following represent some examples in the context of the negotiations discussed above.

Problem 1

Wrong: "You didn't allocate enough money in the budget for what we are expected to do in the proteomics facility. That's why we need this mass spec now."

Better: "Our budget in proteomics doesn't provide sufficient funds for our assignment. As a result we're behind in our workflow."

Problem 2

Wrong: "You do this all the time. You come to us at the last minute and make us drop everything to do the study that you think is most important. My team is really frustrated by this stop/start mentality."

Better: "My group functions better if they can follow through on our projects. I understand that you're under pressure, and I think that we might be able to come up with some alternative approaches to helping you."

Problem 3

Wrong: "You have no appreciation for all of my work on this project. I work overtime and my only reward is more work."

Better: "I feel that the hard work I did on this project has gone unappreciated. I think that I deserve some concrete acknowledgement of the importance of my accomplishments."

Chapter 6 provides additional examples for using this technique in a negotiation.

Learn to expand the pie

Once you learn to negotiate based on interests, and not positions, you are on the road to opening up a spectrum of possibilities and solutions. Look at the interests we identified in the three sample problems earlier in the chapter. Each of these interests might not have been obvious had we not delved beneath the surface of the problem. As a consequence, we were able to extend the scope of the negotiation far beyond what it might have initially focused on. As a scientist, you may tend to pare the problem down to its essential elements: get to the root of the difficulty, simplify as much as possible, and negotiate about the one key element. In negotiation, this is a big mistake.

Let us return to the negotiation involving first authorship of a scientific paper. In its simplest form, the negotiation seems to be a classic win/lose situation: There can only be one first author. It is true that some journals now include footnotes indicating that the first and second authors contributed equally to the work, but many scientists would still opt to be the first author if they could. After all, when your paper is discussed in the "Current News" section of *Nature*, you would rather see it referred to as work by "Your Name et al. 2005" instead of "Someone Else's Name et al. 2005."

> ▸▸ *Case Study: First Authorship*
>
> Alex and Pat are postdocs in Dr. Waffle's lab; Alex is one year senior to Pat and Pat is new in the lab. For the past year, Alex has been working on a project to generate adult stem cells from rodent liver. Although he has made good progress and has established important tools and assays for the work, the project is bogged down. In an attempt to help, Dr. Waffle assigns Pat to work with Alex. Pat has experience in isolating mouse hematopoeitic stem cells and introduces several powerful new approaches to Alex's experimental system. In a very short time, Pat's modifications yield spectacular results and Dr. Waffle instructs the pair to write up a manuscript describing the work. Dr. Waffle tells Pat and Alex that they will be first and second authors, he will be third author, and it is up to them to decide which one will be first and which one second.
>
> Pat and Alex meet several times over the course of the next month to discuss the manuscript. Each time, they have a brief discussion about order of authorship, but it quickly becomes clear that they both believe that they deserve to be first author.
>
> Alex is planning to start looking for a job in six months and believes that first authorship on this seminal paper will be his key to a great job. Pat plans to continue working on the project in Dr. Waffle's lab for at least another year and feels that his contribution was the sole reason that the project got to where it is today. Pat feels that it is simply a matter of fairness that he be first author.
>
> Neither will budge from their position.

If you are a scientist, the interests of the two parties may seem obvious to you: They each want to be first author. But that is not an interest; it is a position. Remember, an interest is what each expects to achieve by becoming first author. A position is simply one way of satisfying an interest. A more detailed analysis of the interests of the two parties reveals a far more complex and useful set of opportunities.

Alex's interests (what he hopes to achieve)

- Professional recognition for the groundbreaking work
- Credit for having started and laid the groundwork for the project
- A good job in the next six months with the help of the publication

Pat's interests

- Professional recognition for groundbreaking work
- Recognition for his specific creative contributions
- Continued work on the project after Alex leaves the lab
- Continued work in this field and eventually a National Institutes of Health (NIH) grant

The first authorship issue represents the tip of an iceberg of underlying interests for each. By identifying these interests, we also identify additional topics for Alex and Pat to negotiate. While this may seem counterproductive at first glance, expanding the pie (adding more elements to the discussion) actually gives Pat and Alex greater flexibility in coming to an agreement. If all or some of the interests listed above become part of the discussion, it no longer seems like the simple win/lose situation of who gets to be first author.

Here are some ways that Alex and Pat can use their interests in an expanded negotiation.

- Both know that a Gordon Research Conference is coming up in three months and one of them will need to present the work. Since one of Alex's interests is in having face-to-face exposure with people in his field who may have job opportunities, presenting the work at the conference might be a good way to accomplish that. In that case, Alex may be willing to agree to be second author if he can present the work at the conference.

- Alternatively, Alex may already have a job lined up and plans to start writing a grant application to the NIH for research support. In this case, being able to continue with the work that is described in the paper would be a great benefit. Thus, he might be willing to let Pat be first author in exchange for an agreement that he will have the right to take the project with him when he leaves.

- It may be that another paper is expected to arise from the collaboration. Perhaps Pat will not look for a job for another 18 months, whereas Alex is looking for one now. Pat may agree to let Alex be first author provided that Pat gets to be first on the next paper and is guaranteed to be able to keep a major part of the project as his own when Alex leaves.

The next two examples of expanding the pie show how inventing new possibilities that benefit both Alex and Pat can facilitate an agreement.

- At Alex's suggestion, Dr. Waffle promises that he can get an invitation from a prestigious journal for Alex and Pat to write a joint review article outlining progress in this field. Both will benefit greatly from the exposure, and neither could write the article on his own. For this article to be written, Alex and Pat need to be on good terms. This is a powerful motivation for them to come to an agreement. Waffle suggests that whomever is second author on the first publication should be first author on the review.

- Alex suggests that in exchange for first authorship, he will send Pat any new cell lines he derives in his new position and that they will collaborate on the work associated with these lines. Pat benefits by having access to these important cells and Alex benefits by having Pat as a collaborator.

The point here is that the more elements you add to the discussion, the greater the likelihood that you and the other person can find a way for your underlying interests to be met. You have expanded the pie by adding additional elements to the negotiation, but you were able to do this only because you were willing to look beyond the narrow confines of the authorship question.

As we noted above, it is important to prepare to expand the pie before you begin the negotiation. Think through your interests and all of the possible ways that they can be satisfied and be prepared to offer new elements to the authorship discussion. In addition, consider ways in which you and the other party can work jointly toward an outcome that neither of you would be able to achieve separately. The last two examples above show how adding such options allows more opportunity to create an equitable agreement.

You can also do this very effectively during the negotiation. By discussing the authorship question in the context of satisfying interests, you turn what could have been a contentious discussion into one of mutual problem solving. By presenting your interests honestly, you become open to ideas and suggestions that may not have occurred to you.

Asking plenty of questions of the other party can help you to expand the pie by identifying the other side's interests that may not have been apparent to you at the outset. For example, you might say, "I'd like to understand why being first author on this particular paper is so important to you. Can you run through that with me?" Although you may think that you already know the answer, you might be surprised if the other person were to say, "I'm leaving research to go to law school and I want to have just one first author paper. You can have the project when I leave and you can present the results at the Gordon Conference."

Alternatively, you could ask, "Is first authorship really the issue for you or are you more interested in establishing a claim to the invention?" or "Hypothetically, if I were first author what would be the downside for you?"

Manage yourself: Know your hot buttons and watch your body language

If you are concentrating exclusively on the content and outcome of a negotiation, without realizing it you may be interacting in a counterproductive manner during the negotiation. For example, you may speak or behave in a threatening way. A scientist in one of my workshops repeatedly jabbed his finger toward the other person's chest when making a point during the negotiation on authorship. How do you think this person felt? That negotiation turned into an argument.

Self-awareness means not only being aware of your feelings and behavior, but also anticipating how you are likely to behave in a particular situation. Most of us have sensitivities to certain things that people do or say. These sensitivities, called hot buttons, may cause us to become defensive or feel angry or hurt. In each case, if we react overtly to those feelings during a negotiation, we may negatively impact an already contentious discussion.

I once worked for someone who was familiar with all of my hot buttons and used them to great advantage for years. Whenever he wanted to win a debate, he would accuse me of being arrogant and selfish. I believed there to be some truth to these accusations and felt guilty about being this way. Moreover, I felt vulnerable and exposed, because I knew that he was aware of these characteristics in me. As a result, every time he accused me, I would become flushed, feel flustered, and lose my ability to continue with our discussion in a calm and dispassionate manner. This often prevented me from thinking clearly and led to my becoming withdrawn. Needless to say, whenever this happened, I lost whatever debate we were having. This type of reaction is what Daniel Goleman has called an "emotional highjacking" (Goleman 1995, pp. 31–51), and that is exactly what it felt like to me.

It took me a long time to learn how to anticipate this highjacking. I started by practicing becoming aware of my feelings during these incidents. This in and of itself was a big step, because simply telling myself that I was feeling angry and humiliated gave me some perspective and control over what I was experiencing. Thinking in this way moved me from a mode in which I was reacting to my feelings to one in which I was thinking about them. Thinking about my feelings eventually enabled me to realize that I could have the feelings without reacting to them at that moment. I was able to notice what was going on inside myself, anticipate behavior that would be counterproductive or harmful to my interests, and decide to behave differently.

At long last, I was eventually able to tell my boss, "I may be a bit arrogant at times but that is not what is at issue here. Let us try to focus on the issues at hand." Managing your hot button reactions requires you to achieve some emotional distance from yourself and your feelings. It is akin to what Ury has called "going to the balcony" (Ury 1993, pp. 31–51), meaning observing yourself and your feelings from a distance. This puts you in the role of an observer rather than a participant. If you can achieve this state of mind and maintain it during negotiation, you can represent your interests more effectively.

See the exercises at the end of Chapter 2 for help in identifying your hot buttons and make note of each as you discover it.

Do not take anything personally during a negotiation

People with low self-awareness are more likely than others to become insulted or take umbrage to things that are said during contentious discussions. However, most of what people say that makes you feel angry or insulted is simply the result of thoughtlessness. Remember, the scientists with whom you work have the same problems as you. If you have a tendency to speak before you think—to say or do things without fully considering their impact on others—you are not alone.

The same behaviors that we excuse in ourselves by saying, "Oh, I really didn't mean to say that; I just wasn't thinking," we attribute in others to malicious intent, a deliberate desire to affront, and malevolence. This is a psychological phenomenon called the fundamental attribution error (Plous 1993, pp. 180–181). Particular behaviors that we observe or experience in others we attribute to fundamental characteristics of their personalities, rather than, for example, careless error, ineptitude, or naïveté.

Even if you think that the other person is deliberately insulting you, ignore it anyway. Remember that your objective during a negotiation is to get your interests satisfied, not to fight. Your best strategy if someone says something insulting is to ignore the comment and continue as you were. Never trade insults.

Of course, if the insults persist and become malicious in character, you can say, "I'm insulted by what you're saying and think that we should postpone this discussion until we can both focus on the matter at hand." Telling the other person how you feel calls attention to the objectionable behavior in a nonaccusatory manner and is often enough to change the person's tune.

The following case study illustrates how a science consultant, Samantha, deals with a difficult client. The case illustrates how a seemingly simple interaction could have turned into a conflict if one participant had reacted differently to an insult.

▸▸ *Case Study: Don't Take the Bait*

The head of a small company that was trying to develop a treatment for Alzheimer's disease contacted me. They had tried to raise money, but were unsuccessful largely because investors were skeptical about their approach. The chief executive officer (CEO) asked me to review their data since I'm an expert in this field. However, a close friend had consulted to this company a year ago and it took the company six months to pay him for his work. He told me that he would never work for them again.

When I spoke to the CEO, I told him that I worked on a contingency fee basis and would need to be paid for the work before I started. He agreed to this and said that he would have a check for me when I arrived. After I arrived, we engaged in small talk, he led me into a conference room, and then launched into his data presentation. I interrupted him and asked if we could get the financial formalities out of the way before we began, as we had agreed. He looked at me a bit askance, but agreed that he would go to his office and get the check.

After about ten minutes, he came back and said that his administrative assistant was out that day and that she was the only one who knew how to print out checks using their enterprise software. He suggested that we proceed and he would send the check later.

I was taken aback by this, especially since we had an explicit agreement that I would be paid in advance. I said, "Well, I think we need to resolve this first. I'm a real stickler for details, and we did agree that I would be paid in advance. If you simply write a check by hand, that would be fine."

He looked surprised and said, "You're really not a very trusting person, are you? Are you being a little paranoid?"

I felt angered and insulted by his comment, and mentally counted to three while I fidgeted with my pen. At that point, after having taken a couple of deep breaths, I said, "It's not a matter of trust, and I'm really not paranoid. But I am a bit compulsive about sticking to the terms of agreements I make. I agree that it may sound strange, but I would be distracted during our discussion and would have a hard time giving you my full attention if we didn't first get this out of the way. So if you would just humor me, we can get on with the meeting."

He shrugged, went to his office, and came back with a hand-written check. The consultation went smoothly from there.

Samantha could easily have taken umbrage to the accusation that she was untrusting and paranoid. She could have said, "We had an agreement and you're not sticking to it, so maybe I should be untrusting. I've heard stories about you... ." Instead, she refused to take the comment about trust as an affront and made light of it, suggesting that although it might seem that she was mistrusting, it was just an idiosyncrasy and she should be humored. At the same time, she was agreeing with the client's observation and empathizing with his feelings, further limiting the likelihood of a confrontation.

Note that Samantha really did feel insulted by the comment, but she managed to recognize her feelings and behave in a manner that enabled a resolution. The key to avoiding taking things personally is not that you should not feel insulted, angry, or hurt. In fact, it is likely that you will never be able to control how you feel in these situations. The key is to give yourself the opportunity to make decisions about how you will behave or respond, regardless of what you are feeling.

Listen to and acknowledge the other side: Be empathetic

How often have you been in a meeting or discussion, and while the other person is talking, you are carefully rehearsing what you are going to say in rebuttal? If you do this, you may hear, "Oh, that is not at all what I said. What I said was.... ."

We often spend less time listening than preparing to refute what we think we hear. This happens often with scientists, because so many of their arguments are carefully reasoned, sequential, and orderly. All of that does take some thinking.

Adele Lynn (Lynn 2002) identifies six types of what she calls "negative listening patterns." These occur when we seem to listen, but in reality we are not listening at all. Below we describe the six patterns, but we modified her list somewhat, for our purposes. Read through the list and ask yourself if you display any of these behaviors when you are supposed to be listening.

1. **Faking.** Make all of the outward signs of listening, i.e., eye contact, saying "uh huh" etc., but do not really listen. The faker is thinking about what they are going to say next or about something else altogether.

2. **Interrupting.** Interrupt the speaker before she is able to finish and/or interrupt with questions or comments that show what you know rather than ask for relevant clarifications.

3. **Intellectualizing.** Become obsessively logical about what the speaker is saying, highlighting and focusing on minor logical flaws and not hearing the meaning or affect.

4. **Free associating.** Pounce on something the speaker says because it reminds you of an association that you need to mention, but that has nothing to do with the content of what the speaker is saying.

5. **Gathering ammunition.** Listen only long enough to hear something with which you disagree or that you can refute, and then pounce on that.

6. **Solving the problem yourself.** Give the speaker advice about how you would solve their problem rather than listen to the full story and understand the speaker's viewpoint.

Can you identify occasions when you have exhibited one or more of these behaviors? Scientists can become obsessed with being right, so they give what they say a lot of thought to ensure that they are not making some error of logic or fact. That's all fine—provided all that thinking and preparing leaves time to listen to what others say. Listening carefully takes much more work than most people think. Remember that in a negotiation not only do you need to listen, you need to demonstrate to the other party that you hear what they are saying, and that takes even more work.

There are many ways to produce a climate conducive to an agreement, but the simplest is to behave in a way that shows the other party that you are listening to and hearing what they are saying. It also helps a great deal if you actually *are* listening to and hearing what they are saying.

Simple methods to convince the speaker that you are listening include repeating what he has just said, "So if I understand you correctly, you're suggesting that ... ." "Ok, so your view is that... ." Other methods include making eye contact with the speaker, and nodding and saying "uh huh" every so often to assure that you are hearing, but not necessarily agreeing with, what she is saying.

You may be surprised to discover that as you act out these listening behaviors, will actually become more attentive and responsive to what the other person is saying. That is, as any behavioral psychologist will tell you, changing your behavior can change your attitude, outlook, and feelings (Burns 1990).

Finally, when listening, take special note of your own reaction to and feelings about what is being said. Pay attention to what might seem like transient thoughts or feelings that lie just beneath the surface of your consciousness. If you can capture these, you may find them extraordinarily useful. Often we intuit something about someone or what they are saying before we are fully conscious of it, and often those intuitions are far more accurate and revealing than what we consciously think.

If you become aware of feelings of discomfort about what someone is saying, this should be a tip-off that you may not be registering their message accurately. This is because your emotions or anxieties may lead you to jump to conclusions, make assumptions, or misinterpret or misunderstand what is being said. In these circumstances, it is a good idea to ask questions to clarify what the other person is saying. Repeating what you just heard and asking if you understood it correctly is a good way to do this.

Defuse anger and hostility

As we have noted, do not respond to or trade insults or accusations during a negotiation. If these reflect anger or hostility, you need to defuse or neutralize the anger before you can engage in a productive negotiation. Chapter 5 presents tools for dealing with an angry boss and these same tools can be used in a negotiation. Jump ahead to Chapter 5 for a preview of how to use the four techniques of agree, empathize, ensure, and inquire. Each of these separately will help defuse anger and, when used sequentially with an angry or hostile negotiating partner, will very likely help you to change the focus from anger to the matter under negotiation.

REFERENCES

Burns D.D. 1990. *The Feeling Good Handbook*. Plume/Penguin, New York.

Fisher R. and Ury W. 1991. *Getting to Yes: Negotiating Agreement without Giving In*, 2nd edition. Penguin Books, New York.

Goleman D. 1995. *Emotional Intelligence*. Bantam Books, New York.

Lynn A.B. 2002. *The Emotional Intelligence Activity Book: 50 Activities for Developing EQ at Work*. Amacom, HRD Press, New York.

Plous S. 1993. *The Psychology of Judgment and Decision Making*. McGraw-Hill, New York.

Ury W. 1993. *Getting Past No: Negotiating Your Way from Confrontation to Cooperation*. Bantam Books, New York.

EXERCISES AND EXPERIMENTS

1 The right tools for each negotiation

Below is a a summary of negotiation tips condensed from Ury's *Getting Past No* (Ury 1993). Make a copy and review it before your next negotiation. You will not be able to apply all of the guidelines to every negotiation, nor should you need to. However, you will find that with certain people, and in certain situations, some of the guidelines will be more important than others. For example, if you are negotiating with someone who is very aggressive, you may need to use techniques to ignore or deflect insults or innuendos. With a very passive person, you may want to use your listening skills to better understand their interests. If a situation seems to be intractable because of the limited number of options available, remember to expand the pie and identify underlying interests.

Negotiation Guidelines*

Keep in mind the following.

- Negotiators are problem solvers, not competitors.
- The goal is a wise outcome, not victory.
- Separate "people" from "problem." Be hard on the problem and soft on the person.
- Focus on interests, not positions.
- Invent options for mutual gain.
- Insist on objective criteria.
- Yield to principle, not pressure.

Before the negotiation, prepare.

- Identify your and the other's interests. Think through options.
- Define standards to be used in reaching an agreement.
- Define what you would like to achieve.
- Define what you would be content with.
- Define what you could live with.

During the negotiation, manage your state of mind.

- When attacked, do not strike back, give in, or break off the negotiation.
- "Go to the balcony" to gain perspective.
- Buy time to think; rephrase what has been said.

Manage their state of mind. Identify with the other side.

- Find something with which to agree and agree whenever possible.
- Paraphrase and ask for corrections to your understanding.
- Acknowledge their feelings.
- Express your views in a nonprovocative manner.
- Make "I," not "you," statements.

Use clarification to identify underlying interests.

- Do not focus on their position; focus on common goals.
- Ask open-ended questions such as "Why?", "Why not?", or "What if?"

Deflect obstacles and negative tactics.

- Focus on fairness, standards, and objective criteria as ways of evaluating an agreement.
- Ignore flat refusals or reinterpret them as aspirations.
- Ignore attacks or reinterpret attacks as assaults on the problem. Never attack back.

*Adapted from Ury (1993)

See the following table and make a list of three or four negotiations that you have had in the past year. Remember as accurately as you can how you felt and acted during each negotiation. Fill in the table, indicating which negotiation tool or concept would have been most useful to you during each negotiation. For each of the negotiations that you listed, check the tools that could have led to a better outcome.

Topic of negotia-tion	Negotiation technique						
	Prepare for negoti-ation	Focus on interests, not positions	Focus on problem, not person	Expand the pie	Manage yourself	Listen to the other side	Defuse anger
1							
2							
3							

2 Identifying underlying interests

Think of a person with whom you find yourself in disagreement on a regular basis. Recall the most recent disagreement or difficult negotiation and answer the following questions (it will be most useful if you write down the answers since you will need to refer to them later in the exercise):

a. What was the disagreement about?

b. What did you hope to achieve? What were the interests or underlying needs that you were trying to satisfy? List as many as you can. Refer to the authorship negotiation case study earlier in the chapter (First Authorship) for examples of underlying interests.

c. For each of your underlying interests, write down one or two ways that you could have added an element to the discussion, besides the primary topic, that would have enabled you to satisfy that interest.

d. What were the underlying interests that the other person was trying to satisfy? (You may need to guess if you do not know enough about this person or their interests.) Ask yourself what this person wants or needs and how these factors might have influenced the negotiation.

e. For each of the other person's underlying interests, write down one or two elements that you could have added to the discussion that would have enabled him to satisfy that interest.

f. What were you feeling during the negotiation? (Refer to the list in Chapter 2; use feeling words, not thinking words.)

g. What did the other person feel during the discussion? What did their body language convey?

h. Review the negotiation guidelines in the summary below. What did you do or say during this negotiation that violated these guidelines? What did you do or say that was in accord with them?

3 Identifying negative listening patterns in yourself

You may be a poor listener without knowing it. Over the next few days, pay close attention to what you are thinking during several discussions with people in your workplace. After each discussion, record in the following chart which of the negative listening patterns that you found yourself using. Note any patterns. Do you use some of the patterns with specific people? Do you use some more than others?

	Negative Listening Behavior					
Situation	Faking	Interrupting	Intellectualizing	Free associating	Gathering ammunition	Solving the problem
1						
2						
3						

4 Correcting negative listening patterns

If you exhibit any of the traits listed in the previous exercise, it is likely that you are missing or misinterpreting what is being said more often than you may think. Test this hypothesis by conducting the following experiment.

Pick three or four people with whom you may use negative listening patterns. Choose a discussion during which you will ask at least six clarifying questions in response to things those people say, *even if you think you know exactly what they mean*. You may find that you are mistaken more often than you think. People use words and phrases to mean different things. Moreover, they are often unclear themselves about what they mean.

• How often did you misunderstand what each person said?

• Were you more likely to misunderstand some people more than others?

• Was the quality of the discussion improved when you began to ask questions?

CHAPTER **4**

A Herd of Cats:
Managing Scientists

Whether you are a molecular biologist or a particle physicist, chances are that you do much of your work in teams. If you manage a team of scientists or technical professionals, you may find that most of your time is spent mediating disputes, ironing out misunderstandings, and placating bruised egos. As we have seen, there is a good reason for this: The people you are managing are focused on technical and quantitative aspects of their jobs, at the expense of the interpersonal and social aspects.

But there may be another reason that you spend a lot of time on these matters: You may be laboring under the same interpersonal deficits as those you manage. This chapter shows you the kinds of problems to expect when managing groups of technically oriented professionals, as well as how to deal with them should they arise. You will also learn about your possible blind spots that can make your job as a scientist/manager harder

than it has to be. Our approach to helping you to become a better manager of scientists starts with helping you to be a better observer of yourself and others.

A GROWING AWARENESS OF TEAMS IN SCIENCE

During the past 50 years, science has increasingly been done by groups of scientists with complementary or overlapping skills. This is especially true in the private sector in companies of all sizes. Although the era of the scientist as individual practitioner is by no means past, it is clear that more and more of the dollars spent on scientific research (both in the public and private sectors) are spent on projects involving groups or teams of scientists.

There is increasing awareness of the trend that scientific endeavors are becoming more complex, interactive, and social. A recent NIH conference on "Catalyzing Team Science"* speaks to the importance of teamwork in the life sciences. Increasingly, graduate programs attempt to impart communication and other "meta" scientific skills to their students.

To say that science has become a social occupation is not to say that scientists themselves have become social creatures, or even that they should if they are not already. Increasing the size of groups adds scientific skills, expertise, and sometimes simply more hands to do the work. In practice, however, the accessibility of members' skills and information to the group depends on how well the member scientists relate to one another. In the worst cases, information and expertise are shared selectively or not at all, data are hoarded like a scarce currency, and team members lie in wait for the most public opportunities to demonstrate superior knowledge. More often than not, science managers and leaders fail to recognize or deal with such behaviors, much to the detriment of the group or organization.

Anyone who has managed science and technical professionals working on complex projects requiring collaboration and interaction knows how difficult this can be. Let us examine some of the challenges and some of the ways of meeting them.

SCIENTISTS MANAGING SCIENTISTS: CHALLENGES AND OPPORTUNITIES

If you are a technical professional, it would probably not come as a big surprise to learn that some people in your organization think that you are hard to manage or a poor manager. In the private sector, scientists are often sent to management training seminars. These typically teach participants to set goals and objectives, give feedback, do evaluations, and manage projects. These are all important skills and worth learning. However, your success at applying these skills is not determined by how well you know them or even how long you use them. It is determined by how well you understand yourself, and how well you relate to and respond to the people to whom you need to apply them. If you are oblivious to your own motivations and feelings, you probably do not pay attention to or understand the motivations and feelings of those you manage. If you interpret silence as agreement, repeated absences as laziness, and failure to follow instructions as forgetfulness, you cannot be an effective manager.

Some of the studies we cited in Chapter 1 suggest why science professionals make such misattributions: They may not notice interpersonal conflicts; discern underlying motives, needs, and expectations; or listen carefully. We would also add that they are probably not very self-aware.

The good news is that as a scientist, you are the best possible choice for managing other scientists, either as a team leader or an executive. In the following we introduce skills and concepts that will improve your ability to manage teams. Case studies illustrate how improving self- and interpersonal awareness can help in real world situations.

*Conference BECON 2003 Symposium on catalyzing team science. June 23–24, 2003, Natcher Conference Center, National Institutes of Health, Bethesda, Maryland (http://www.becon.nih.gov/symposia_2003/becon2003_symposium_final.doc).

Ignoring problems and conflict in a team

The interpersonal difficulties of scientists often stem from an aversion to admitting that a problem exists. If you are unsure what to do about a problem, or are uncomfortable thinking about it, you will likely avoid it. But if you are a team or group leader, ignoring problems in your group can have a detrimental effect on team morale and productivity.

Some common problems with which scientific team leaders have difficulty dealing include

- Employee performance or attendance problems. Perhaps your technician consistently has difficulty with crucial experiments, keeps undecipherable notes, or routinely misses one or more days of work each week. You compensate by double-checking his protocols, having him e-mail his raw data so that you can analyze it, and rescheduling experiments because he is absent. What you do not do is address the problem with the technician.

- One member of your group consistently complains about the lab, other people, and every piece of equipment that he uses. He also finds fault with every lab policy and voices these complaints to anyone who will listen. People in your group tell you that his negative attitude is getting on everyone's nerves. Your response is that his complaints are harmless and should be ignored.

- A new member of your lab has personal hygiene habits that others find offensive and distracting. He bathes infrequently, which makes working near him unpleasant. When a lab member complains about this to you, you are sympathetic, but find it impossible to imagine how to broach the topic. The person becomes a pariah in the lab and wonders why no one talks to him. His work, and yours, suffers because no one gives him the advice or assistance that he needs.

- One of your technicians talks on her cell phone using a head set while doing experiments. She speaks in a loud voice that you can hear in your office across the hall. Although no one mentions this to you, you suspect that others in the lab are having difficulty concentrating. You decide to ignore it until someone complains.

- One of your most capable scientists is a bully who routinely manipulates others into relinquishing equipment time, technician help, and supplies. He behaves as though his work is more important than theirs. You do nothing because he is productive and you figure that the group members need to work it out among themselves.

- A female employee in your group has a hard time getting her point across in meetings because she is soft-spoken and gets routinely interrupted by several loud males. You observe this but decide to do nothing, figuring that she needs to learn how to roll up her sleeves and jump into the fray if she is going to succeed.

- A female employee complains to you that a male co-worker is viewing pornographic images on his laptop in the conference room. You are at a loss about how to broach the subject to him, so you do nothing.

These situations, to which we return shortly, are all examples of conflict avoidance, one of the most common and damaging mistakes made by managers of science teams. Recall that lack of awareness of conflict among team members was one of the personality characteristics that Gemmill and Wilemon (1997) found to be more pronounced in scientists than in nonscientists. The simplest way to avoid conflict is to be oblivious to it. If you do not pay much attention to other people, miss subtle cues in their behavior or manner, and if you ignore, dismiss, or trivialize what makes you uncomfortable, you will not be aware of conflicts simmering all around you.

If the conflict erupts into a full-fledged war, as happens in the case study below (Ignoring Conflict), it may be impossible to ignore. In this case, you may take another path to avoid conflict by simply doing nothing about it. I have heard many scientists in responsible positions assert that the reason that they do not intervene in whatever conflict is under discussion is that they think the warring parties ought to "work it out themselves." But if you have gotten this far in the book, you may suspect that this is code for "I don't have a clue to how to help resolve this problem." This is the main reason we avoid conflict: We do not know how to deal with it. Moreover, many feel that conflict is to be avoided at all costs—it should never occur in the first place.

This, of course, is all wrong. First, conflict is inevitable and can even be a useful mechanism for bringing out differing views. Second, working through a conflict does not have to involve angry confrontation, insult, or accusation. Third, you can learn to work through conflict in a collegial and productive manner. Many of the negotiation techniques that we introduced in Chapter 3 are also good tools for resolving conflict. If you suspect that you ignore or avoid important issues because you are conflict-averse, try one or two of the tools in Chapter 3. But take your time. If you have spent your life ignoring conflict, do not insert yourself into thorny situations all at once.

The following case focuses on a team leader who is unaware of a simmering conflict in his lab, with unfortunate consequences for the project and the group's productivity.

▸▸ *Case Study: Ignoring Conflict*

Ralph was a senior environmental policy analyst in the Environmental Protection Agency (EPA) and he ran a group specializing in industrial ground water contamination. He was preparing to submit a lengthy and technically detailed report on behalf of the agency regarding a ground water contamination suit being reviewed by a state court.

While working on the report, he noticed some discomfort on the part of two of his three junior associates, Richard and Teresa, about what exactly should be included in the report, but he did not ask them about it. He finished a close-to-final draft and gave it to all three junior associates for review.

Richard and Teresa objected to including the third associate, Tony, as an author of the report. Ralph explained that Tony had contributed some of the analysis cited and should be named. Richard and Teresa said that they did not trust Tony's results and demanded that Ralph omit Tony's name from the report. Ralph was stunned. He had no reason to believe that Tony's analysis was suspect, but he did notice animosity among the three associates.

Ralph responded that Tony had been part of the team all along and was deeply involved in its planning, analysis, and strategy. He concluded that the project had been a team effort and that Tony's name should remain. Richard and Teresa refused to have their names included on the report unless Tony's name was removed. Ralph was anxious to resolve the situation before the end of the day, which was the latest that the report could be FedExed to reach the court in time. He pushed the discussions, but Richard and Teresa grew more and more agitated. Ralph called their behavior outrageous and likened it to blackmail. At that point, Richard stormed out of the office and did not return until the following day, after the deadline had passed. The EPA lost its chance to present the work to the court and Ralph started disciplinary action against Richard.

There are many ways of looking at this situation. One is to take an approach that might come naturally to technically minded people, that of seeking the truth. What were the facts? Did Tony's name actually belong on the report? How much did he contribute? Was his work in fact shoddy? Is there some set of rules by which these facts could have been ascertained and weighted? These are all relevant questions, but I suspect that answering them would not have solved Ralph's problem. The real problem was that Ralph had not recognized and addressed the interpersonal conflict that blew up in his face. If Ralph suspected some animosity between Tony and the others, he needed to address it. Ralph explained why did he not address the problem earlier:

> "I noticed some discomfort among these people, but I had never had any trouble with Tony, and I couldn't see why Richard and Teresa objected to what he was doing. Richard and Teresa never said anything critical of Tony's work in our group meetings, and frankly, I thought that there was some personal reason that they didn't like him."

The result was that the report was filed too late to have an impact on the court case, Richard was disciplined and eventually left the agency, and the entire group, especially Ralph, was viewed as incapable of meeting deadlines. Everyone lost.

What would have happened if Ralph was aware of his tendency to avoid and ignore conflict? He might have made a special effort to be on the alert for signs of animosity in his group, if only because he knew that he often missed the cues. In this case, he might have noticed that there was a problem, tried to determine what was bothering Richard and Teresa, and done something about it. Perhaps they had misinterpreted something Tony had said or done. Perhaps Ralph would have discovered that Tony's work really was shoddy. What if Richard and Teresa had been able to separate their personal animosity toward Tony from their professional concerns and discussed these concerns with Ralph? What if Richard had been able to control his temper and enumerated his objections in a way that Ralph could understand?

If even one of the participants had had more insight, self-awareness, or ability to handle conflict, the outcome would have been different for everyone. Even a single individual with good self-awareness and interpersonal skills can have a profound impact on how a group functions. It is easy to focus on Ralph as the one responsible for the debacle. If Ralph had been more attuned to the interactions of his team, and if he had confronted them with observations that team interactions were deteriorating, this incident might never have happened. However, Ralph is not solely to blame. Each of the others had responsibility for their own role and inability to deal with the conflict.

If you are not a government employee, before you start feeling smug about the ineptitude of narrow-minded civil servants, try this experiment. Reread the above case study and change the participants from a team of EPA analysts to a team of engineers in dispute over the design of a Mars rover, a group of middle managers developing a plan for a new business unit, or a group from any technical discipline in which you work entrusted with reaching a goal. My guess is that you will find a lot that feels familiar in the example regardless of the type of work you do.

The right words

Because finding the right words to say in a conflictual situation is so difficult, we offer below some specific suggestions. Notice in these examples that we are following the advice of Chapter 3 to focus the discussion on the problem and its consequences, rather than on what the other person did or said.

The following suggestions are framed in the context of a conversation with another team member or colleague about something that you see as a potential conflict.

1. Start your sentence with "We have a problem." Starting with "we" immediately frames the problem as being shared, not the fault of one person.

2. State the problem, being as concrete as possible. For example, "We aren't communicating effectively. There seems to be confusion and misunderstanding about what I expect and what you think I expect." Then fill in the specifics of the situation.

3. Focus your comments on how the problem impacts shared objectives. This depersonalizes the difficulty, making it easier for the other person to take what you say as constructive criticism.

4. Make sure that you both understand what the other is saying. Ask if anything you said is unclear, and repeat what you heard to confirm that you understand what they said ("I want to make sure that I understand your point of view. Here is what I am hearing..."). Ask the other person their version of the problem. Ask for clarification frequently ("What do you mean by...?"). Spend as much time as needed explaining or asking questions until you are sure that you are both discussing the same problem. If you skip this step and move to solutions too quickly, you may each be solving different problems.

5. Often this process itself leads to solutions. At this point, you are not pushing for them, but they may just begin to present themselves. Defining the problem illuminates things that may not have been visible before. As the solutions come, write them down, but do not evaluate them at this point.

6. Try generating solutions to the problem. Start by focusing on yourself and what you can do to improve the situation. Then ask what the other person can do. If she does not suggest anything, offer suggestions yourself. Keep the focus on solving the problem and improving the situation, not on changing the other person.

7. Continue to exercise your self-awareness and communication skills during the process.

Let us apply these guidelines to five of the examples cited at the beginning of the preceding section.

Example 1

Your technician consistently has difficulty with crucial experiments, keeps undecipherable notes, or routinely misses one or more days of work each week. You compensate by double-checking his protocols, having him e-mail his raw data so that you can analyze it, and rescheduling experiments because he is absent. What you do not do is address the problem with the technician.

Start by making a list of several specific instances of the behavior in question. Frame the list in terms of how it has affected the work, the project, or the lab. A hypothetical conversation follows:

You: "Jim, I'm having this conversation with you because your work isn't as good as it could be. I've noticed that your attention to detail and your attendance have declined recently but I know that you can do better. Is there anything affecting your work that I should know about? Last week you forgot to add buffer to half of the sample tubes, and the week before you left the samples incubating twice as long as needed. As you know, we had to repeat those experiments, and they're costly."

Jim: "I just think that all of the scut work gets dumped on me."

You: "Jim, what you do is every bit as important to me as anyone else's work. What I'd like is to find a way to help you do it better. Let's go over several areas where we need to work on improving your performance. I'll help you with this as much as you need because I value your work. Let's start with attention to detail. During the next month, we'll have a number of critical experiments in which you will have a crucial role. I'd like to set a goal for no experimental errors during the next several months. Can you commit to that?"

During the next month, Jim's performance does not improve. In fact, it deteriorates. You do not ignore the problem. You call him into your office and have the following conversation with him.

You: "Jim, I'd like to review your performance during the past month with you. Specifically, we need to talk about your absentee rate and your attention to detail in experiments 34 and 35. These are the same issues that we've been working on for the past month. Are you having any problems that might be impacting your work? If not, I need to tell you that if these areas don't improve during the next month, we'll have to discuss whether this is the right job for you. Can we go over the specifics now and try to figure out the problem?"

Example 2

One member of your group consistently complains about the lab, other people, and every piece of equipment that she uses. She also finds fault with every lab policy and voices these complaints to anyone who will listen. People in your group complain that her negative attitude is getting on everyone's nerves. Your response is that her complaints are harmless and should be ignored.

You: "Melanie, do you have some problems in the way the lab is running that you'd like to discuss with me now?"

Melanie: "I don't have any problems. What do you mean?"

You: "I know that you've expressed dissatisfaction with the way people are assigned to maintain equipment, as well as the allocation of travel funds. You may have some legitimate concerns, but I can't deal with them unless you talk directly to me. I'd like your commitment that the next time you find something that can be improved in the lab, you'll come directly to me. I promise that I will listen carefully to what you have to say."

Example 3

A new member of your lab has personal hygiene habits that others find uncomfortable and distracting. He bathes infrequently, which makes working near him unpleasant. When a lab member complains about this to you, you are sympathetic, but find it impossible to imagine how you can broach the topic. The person becomes a pariah in the lab and wonders why no one talks to him. His work, and yours, suffers because no one gives him the advice or assistance that he needs.

You: "Alan, this is an awkward conversation for me to have with you, but I think that you'll be thankful that we spoke when we're done. Everyone has different personal habits, and for the most part, these are their own business. But once in a while those habits interfere with other people unintentionally. In your case, I've noticed that you don't seem to wear a deodorant. Although that is your personal business, and there may even be a health reason, I need to mention this because, frankly, your "scent," if I can call it that, really distracts me and, I suspect, others as well. Is this something of which you are aware?"

Alan: "No one ever mentioned that to me. I don't wear deodorant because it seems unnatural. People should smell like people, not perfume counters."

You: "I respect your view. But in this case, we also need to consider that you work in close proximity to others who probably find it hard to share the lab with you. I don't think that this is healthy for the lab and it's not great for your relations with the lab members. Do you have any thoughts about what we could do?"

Alan: "Well, I guess I could shower every day before I come to work."

You: "Terrific. Please try that and let's see how it works out."

Example 4

One of your technicians talks on her cell phone using a head set while doing experiments. She speaks in a loud voice that you can hear all the way into your office across the hall. Although no one mentions this to you, you suspect that others in the lab are finding it hard to concentrate. But you decide to ignore it until someone complains.

You: "Natasha, I don't know if you're aware of this, but I can hear you on your cell phone all the way into my office. I suspect that others in the lab may be bothered by this, although no one has said anything to me about it. In the past, I have asked people in the lab not to play music that others can hear and to keep their personal phone calls to a minimum. Is there some particular reason that you need to talk on the phone while you work? Are you dealing with any problems?"

Natasha: "No, there's no problem. I just think that it saves time to talk and work at the same time. That way, I don't have to take time away from my work."

You: "Well, I wish everyone were as concerned about maximizing their time at the bench! Nonetheless, although your talking this way may enhance your productivity, I fear that it will decrease others', including mine. I don't like to make rigid rules for the lab, but I'm going to ask you to restrict your cell phone conversations to the lunch room.

I'll make sure everyone on the team knows about this new rule, and in a few weeks we can discuss how you are doing."

Example 5

Sandrine, a female employee in your group, has a hard time getting her point across in meetings because she is soft-spoken and considered in her speech and gets routinely interrupted by several loud males. You observe this, but decide to do nothing, figuring that she needs to learn how to roll up her sleeves and jump into the fray if she is going to succeed.

Here, you have the option of either telling the interrupters to pipe down or helping Sandrine to become more insistent on getting air time. You may decide on a bit of both:

You: "Sandrine, I've noticed that you often have a hard time getting a word in at team meetings. I wonder why you don't stand up to Fred when he interrupts you."

Sandrine: "I just don't like to argue with him. He just keeps talking louder and louder. I prefer to just keep quiet until he calms down."

You: "If you're willing, I'd like to help you learn to assert yourself a bit more in those situations. It's important for both the lab and you that you get the opportunity to express your views in our meetings. You know a lot about what we are working on and the other team members need your input. Are you willing to try?"

Sandrine: "Sure, I'll try anything that may help."

You: "The next time that you are interrupted, try saying, 'Excuse me, I haven't finished.' If Fred or others interrupt again, say, 'You'll get a chance to respond as soon as I finish' or 'I'd like to hear what you have to say, Fred, as soon as I'm finished.' You might say, 'If you keep interrupting, it's just going to take me longer to get to my point.' If you do this consistently, Fred will get the point. Be patient and don't attack or insult him; keep focused on saying what you have to say. Are you comfortable with these suggestions?"

Sandrine: "Yes, they sound great. Thanks."

You can also send clear messages to your group that suggest that although lively discussion is important and stimulating, rude interruptions are inappropriate. In the long term, your own behavior provides a model. If you interrupt while others are speaking, or shout over them when they are talking, your group will likely feel free to do the same.

Example 6

A female employee complains to you that Juan, a male co-worker, is viewing pornographic images on his laptop in the conference room. You are at a loss about how to broach the subject to him, so you do nothing.

You know that you cannot ignore this problem, but no matter how hard you think about it, you cannot find a way to approach Juan about this accusation. You decide to seek the advice of Harriet, a friend in the human resources department. In a series of meetings, Harriet explains the institution's policy on the use of computers at work and you draft a more specific policy about personal use of computers in the lab. After you announce the new policy to your lab, you hear no further complaints about Juan.

The key in this case was in seeking help from a friend in human resources. If you do not have a friend in human resources, make one. One of the most important tasks of a human resources professional is to help managers find solutions to employee-related problems.

The preceding tools and concepts should help you to recognize and deal with some of the issues that science teams face. We have emphasized how self-awareness on the part of the leader can help you to sense difficulties that both you and your team may be experiencing. We also introduced the notion that both team leaders and members must be comfortable dealing with conflict, which is inevitable in teams. When conflict goes unrecognized and unaddressed, the best-case result is lost opportunity, and the worst case is project derailment or failure.

Technical turf wars

The following hypothetical case study involves a conflict that was both overt and destructive to the team and its progress. Despite being readily apparent, the conflict was handled poorly by all involved. In the sections that follow, we use this case to illustrate and address other problems in managing teams of scientists.

▸▸ Case Study: Technical Turf Wars

One of my jobs in the semiconductor industry thrust me into the middle of a multimillion dollar collaboration between my small research company, Monotech, and a high-profile electronics company, which I will call BigTech. On my first day, I was sent to observe one of the weekly project review meetings attended by teams from both companies. I felt a sense of eager anticipation. The meeting was in one of BigTech's elegant conference rooms with a commanding view of an urban river. As I settled into a seat that had more levers, buttons, and adjustments than I had ever seen on a chair, I had a great view of sailboats taking advantage of the last good days of fall. My enjoyment evaporated quickly: Almost as soon as the meeting started, everyone on my company's team, including Andrew, the project manager, started shouting. It was a scientific free-for-all and the invective was almost unbelievably intense and hostile. I actually felt nauseous. I kept thinking that I was watching a train wreck in progress. In my role as vice president of Monotech, I was ultimately responsible for this impending disaster—and I had just started work that day. What had I gotten myself into? How had this project deteriorated to this point?

Andrew was an experienced individual with a Ph.D. in electrical engineering who had been specifically recruited to manage the collaborative project with BigTech. On paper, his background and experience suggested that he was well suited to managing this project. Moreover, the project was actually very promising and could result in a very profitable product. But the project was moving very slowly because of technical problems.

The agreement between Monotech and BigTech required that our senior project scientists meet once every week to review progress. These meetings (and the one that I had seen that first day was the rule, not the exception) were like rugby matches with players from both sides having a scientific brawl in the mud. Having worked in science for almost twenty-five years, I was used to a certain amount of hostile posturing by aggressive scientists. What I was not prepared for was that much of the mud slinging and invective was among members of my own team! As if this were not disconcerting enough, even Andrew himself lost his cool more than once, and routinely turned red with annoyance at his colleagues.

Back at MonoTech, Andrew complained about his team to anyone who would listen, especially about the fact that members of his team rarely listened to him—even though he was the manager. It did not take me long to see a multitude of causes for this disastrous situation. First, the scientists felt that they were the experts in this particular area, not Andrew. Andrew therefore had no scientific credibility with them. On several occasions, Andrew had presented Monotech's data to the joint project team; this made the scientists furious.

Second, Monotech's internal project team meetings invariably degenerated into accusations of incompetence or worse. During these meetings, Andrew himself became furious and lost his capacity to continue with the discussion. More than one meeting ended with a crimson-faced Andrew ready to explode and a smug group of scientists congratulating each other for so effectively pushing his buttons.

Neither Andrew nor the scientists had the skills to resolve the conflict within which they found themselves. I suspected that the scientists were using Andrew to vent their frustrations with a technically difficult project. For his part, Andrew could not see that by losing his temper, he was just exacerbating the situation. Monotech was also responsible: They had made Andrew project manager because he had a history of managing big projects in other companies, but what they failed to predict was that his limited scientific background in the specific area of the project would be a red flag to the headstrong scientists, who had already burned out two previous managers. Andrew's limited self-awareness, self-control, and people skills simply fueled the flames.

To make matters worse, Monotech badly needed this project and the associated funding that BigTech provided. Everyone knew this, and they also knew that the project was not going terribly well. Yet Monotech's senior managers never openly discussed with the scientists the consequences if the project were cancelled. As a result, there was considerable anxiety, uncertainty, and wild speculation among the scientific staff. The outcome was a disaster with multiple causative factors. Not long after I joined the company, BigTech cancelled the project.

This case study illustrates a series of errors in managing science teams—some of commission and some of omission. Andrew is not the only one who contributed to the problem; each of the participants as well as the senior managers of the company had a role. Some of the problems in this case include the following:

- Management failed to recognize that this group needed a project manager with scientific credibility who would be immune to the hostilities of the group.

- Andrew lacked self-awareness and self-control. This made it impossible for him to weather challenges to his authority and keep the group focused on task.

- The team had a propensity for channeling their frustration into hostility toward one another. Management never intervened to stop this.

- Andrew usurped the scientists' data during project presentations, cheating them of the opportunity to get credit for their own accomplishments.

- Management failed to be open with the scientists about the status of the collaboration, and what might happen if it were to be terminated.

How can disasters like the Monotech case be avoided? In the following sections we review the causes and consequences of the team-based problems we identified in the Monotech case. Building on the information we presented in Chapter 1 regarding psychological characteristics of scientists, as well as our suggestions for improving self-awareness in Chapter 2, we show how you can minimize the chances of these and other common problems from derailing your team.

Providing individual recognition in a team-based project

In the Monotech case, Andrew's failure to champion his member's contributions and his usurpation of their results when he presented their data were two of many errors that contributed to the chaos of that team.

Explicitly acknowledging the individual contributions of team members can provide much-needed recognition, even when they do not own a specific element of the project. Going out of your way to notice and mention the roles of each of the team members counts a lot, and can be the single most important contribution you can make to the team's performance.

According to Katzenbach and Smith (1993), "A team is a small number of people with complementary skills who are committed to a common purpose, performance goals and approach for which they hold themselves mutually accountable." In addition, we often think that team members benefit from mutual support and are in some way bonded or connected to each other, if only through shared objectives. But we believe that several elements and implicit assumptions in this view of teams are at odds with the reality of teams in the scientific workplace. First is the tension between the team being the primary work unit of an organization and the scientist's need for individual recognition. Second, and closely related, is the dichotomy between the need to sublimate individual needs and interests to the team's objectives and the need to own some part of the project that can be identified as one's own.

Scientists are typically individual contributors, not team players. Recall that in Chapter 1 we identified autonomy and independence as two characteristics prevalent in scientists. As one pharmaceutical company scientist interviewed by McAuley et al. (2000) explained it, "The scientific side is a collection of one-man bands who amazingly get things done." In my experience, the principal challenge in managing teams of scientists lies in the genuine need of creative scientists to obtain individual recognition for their contribution to the work of the team.

Indeed, if you do not provide opportunities by which the achievements of individual team members can be identified and recognized, the members may attempt to create the opportunities themselves. They may try to design a subproject of their own or initiate an experimental approach that has not been discussed. While such activities can be valuable to the project, they may also subvert or distract the team from its task. If scientists go off on tangents (no matter how creative or inventive) as a way of showcasing their independence and resourcefulness, they are probably not focusing on their primary responsibilities. The team leader must encourage creativity, independence, and inventiveness while balancing these with the need to maintain focus on assigned tasks.

As noted by Katzenbach and Smith, "Real teams always find ways for each individual to contribute and thereby gain distinction." (Katzenbach and Smith 1993, p. 14.) If the authors' use of the word "real" was meant to convey the notion that most teams find it hard to accommodate this ideal, we are in full agreement. Yet this is probably the single most important concept for the leader of a scientific team to grasp.

If the team grows larger and members' tasks begin to overlap or become shared, it becomes increasingly difficult for individual members to own unique domains of responsibility. Moreover, as teams and projects become more complex, more work may get done in the background. Individual contributions to the project may become secondary or tertiary in nature: A technician figures out why the cells growing in the lab are dying and fixes the problem; a colleague helps a team member get urgently needed results on a mass spectrometer. These are important contributions that take time and creativity on the part of the participants, so it is a serious mistake not to acknowledge these to the individuals and the group.

Balancing task ownership with team participation

Your role as a team leader is to ensure that the owners of specific tasks are recognized for their role in the project and are acknowledged and rewarded for their contributions, no matter how small. Leaders who recognize this create teams in which each member has a clearly defined and preferably nonoverlapping function. This makes ownership and identification of individual contributions transparent. Moreover, it enables assignment of responsibility and credit. A team of technical professionals needs to be no more than a group of individuals who bring disparate skills and input to a project. The only one who truly owns the team's objectives is the project manager or team leader. The members of the team each own tasks for which they will be held individually accountable and for which they seek and expect individual recognition.

Although each team member may operate as an individual contributor, the group can work together to integrate information, test ideas, and, in the most creative groups, catalyze new ideas. When the group cannot work on these areas jointly, they can be done by individuals or the leader. In some cases, the work must be done by people outside the group. If you are a team leader and your group has a high level of cohesion, sharing, and collaboration, continue doing whatever you are doing. If your team does not have these characteristics, but the work of the team is getting done, there may be no need for concern.

Rather than try to develop camaraderie, focus instead on modeling respect and tolerance among team members. Team leaders who put down or devalue others or do not acknowledge their contributions, but who go out for a beer once a week with the lab to promote camaraderie, are only fooling themselves and shortchanging their team members. Do not fall into the trap of believing that just because *you*—the team leader—see the group as "all for one and one for all" that everyone else does, too. Although teams can and do accomplish more than individuals, and members can and do derive satisfaction from group achievement, do not forget that scientists need ownership of and responsibility for well-defined accomplishments.

Creating more effective teams, one member at a time

The team concept is perhaps one of the most overworked themes in management literature. Management pundits espouse the view that the secret to success lies in inducing team members to work selflessly toward a common goal. Some of the approaches used to foster such attitudes work well in the corporate world. But we believe that most of what passes as common wisdom about managing teams cannot be applied in the science workplace.

From what we have said above, creating effective teams in science may be more a matter of meeting the needs of the individual members than of creating a shared vision or an all-for-one and one-for-all attitude that so many team-building courses and exercises promote. Some of these needs can be met if leaders provide clear ownership of tasks and recognition of accomplishments. But the most important tools for creating effective teams are the skills that enable team members to function effectively in a team setting. These include the ability to manage themselves in contentious and stressful situations, and recognize and resolve conflict in a productive manner. These are precisely the skills that the Monotech scientists lacked and their leaders failed to either model or provide.

Christopher Avery (Avery 2001), in his book *Teamwork Is an Individual Skill,* suggests that the most important learning for successful teamwork takes place at the individual level, and not at the level of the group as a whole. Since science teams frequently lack cohesion and a sense of shared ownership of goals, optimal team performance occurs when each member takes responsibility for his own part of the project as well as his individual behavior and performance.

Helping your team to see the big picture

As a team leader, one of your most important roles is to help your team to see the overall mission and goals. This can take several forms.

Interpreter of internal events

Scientists may be myopically focused on their specific project, but have both a need and desire to relate what they are doing to the big picture. It does not take much for a leader to routinely update the team on how their project is related to and impacts the goals of the division or the organization as a whole.

New team leaders often mistakenly assume that because they themselves see the big picture, and the relationship of their project to the organization as a whole, everyone else on the team does as well. Conversely, they may assume that because scientists are so focused on science they do not care about the larger objectives. Both are false assumptions that result in team members feeling disconnected from the company's or organization's goals, and not understanding how their tasks relate to those objectives. Leaders of science teams in the private sector need to pay special attention to helping the team relate their project to the company's overall mission and goals.

A team leader must serve as a buffer between decisions made by senior management and their team members. As discussed in Chapter 7, science professionals become very committed to and identified with their projects. Having a project terminated for lack of progress or—far worse—for business reasons that have nothing to do with the how well the project is progressing can be a frustrating and disillusioning experience for young professionals. This is what happened in the Monotech case. Although the scientists feared that the project might be terminated, they were unprepared when it happened, and had little understanding of the reasons behind the termination. It is the leader's responsibility to ensure that team members understand the reasons for such decisions. Andrew could have played an important role in this process by serving as a conduit between senior management and his team members.

Often, team leaders who themselves are bitter or disillusioned about terminated projects will share these feelings with team members under the mistaken impression that this creates a bond with the team, or that honesty is the best policy. In fact, when the team leader behaves in this way, the team's sense of alienation will grow and it will be more difficult for them to commit to future projects.

Interpreter of external events

In your role as buffer for the team, you serve as interpreter of external events and the business or funding environment in which the organization functions. This helps the team to manage and cope with the uncertainty and ambiguities that can arise from sources outside the organization. Members may feel uncertainty relating to the organization: Will the company be acquired by a competitor? About external events: Will the Air Force buy the company's new guidance system? Will changes in the NIH budget impact funding prospects? And about the industry: Why is the biotechnology stock index down 30% for the year? The team leader can have a pivotal role in providing factual information and helping the team to come to grips with the uncertainties they engender.

One of the biggest challenges that teams of technical professionals face is instability in their field of work or organization. This has been an especially vexing problem in the life sciences and the volatile biotechnology sector, but also common in start-ups in general and in many areas of endeavor. Unstable organizations may negatively impact commitment and encourage an "every man for himself" mindset. The team leader can mitigate the impact of such a climate by helping team members to understand and discuss the underlying issues and how the organization can respond to them.

Helping your team to manage uncertainty

Good leaders address discomfort, distress, and ambiguity in the group and in its members. This means that leaders must be open to signs of distress in their team and actively help the team manage ambiguities and uncertainties. Just as important, leaders need to recognize these feelings in themselves.

If you are feeling anxious and uncertain, chances are that your team members feel the same way. By becoming adept at sensing and identifying these feeling and emotions in yourself, you can use them as sentinels to alert you to incipient problems that need attention.

In his book *Connect*, Edward Hallowell suggests, "Think of these emotions as noises in your car engine. Investigate them." (Hallowell 1999, p. 123.) Treat your own feelings and reactions as data that alert you to situations requiring attention. Self-awareness is just as important for managing others as it is for managing yourself.

When you do sense concern about the future, of either the project or organization, go out of your way to help team members see the big picture. Hallowell points out that because creative professionals often keep to themselves and may be reluctant to share their thoughts and feelings, concerns and fears may become magnified and exaggerated in their minds. He advises, "Get the facts. Very often stress and worry emerge from the imagination, not from reality, particularly for creative people working alone." (Hallowell 1999, p. 123.) We would add that even when fears have a basis in reality, we often exaggerate the consequences. An effective leader promotes communication and provides a reality check for the team.

As a manager of scientists, you must give your group a realistic picture of its future. This is especially important for young professionals whose only working experience has been in academia, where projects can last for years at the discretion of the lab or project director. Even though young professionals in the for-profit sector may claim that they understand that industry norms differ, they may still react in a bitter or disillusioned manner when unexpected changes in projects, leadership, or organization occurs.

Helping your team with uncertainty can take very concrete forms and have important benefits. A senior executive in a major Midwestern company told me that he went out of his way to post a list of projects waiting to be tackled as soon as the current projects were completed. He explained, "My people used to get very nervous when it looked like we were going to deep-six a project. Many of them really thought that if we killed the project, they would lose their job because there would be nothing for them to work on. This had the disastrous consequence that they would perform all kinds of technical acrobatics to keep a demonstrably bad project alive for as long as they could. Posting this list made a tremendous difference because everyone in the organization could see that we had a large pool of exciting projects waiting in the wings. Now, bad projects get killed much earlier."

Of course, in some cases, the leader cannot provide unambiguous answers to the questions that trouble team members the most. In the Monotech case, Andrew need not, nor could he, have given the team solutions for their problems or answers to their questions. But by openly discussing the possibility of project cancellation as well as its consequences, he could have helped them to manage their uncertainty. He could have said, "I'm feeling a bit uncomfortable about the way BigTech executives are talking about this program, and wonder if the rest of you have picked up on that" or "I've noticed that some of you seem to be anxious about what happens if this project is cancelled. Is that true?" The process of discussing these feelings with the group can create a sense of psychological relief by showing them that they have common concerns. If this results in even a small decrease in anxiety and preoccupation with dire outcomes, the work of the team will improve.

Ronald Heifetz, in his insightful book *Leadership Without Easy Answers* (Heifetz 1994, p. 110 ff.), argues that one of the most important attributes of a leader is the management of uncertainty. Leaders do this by creating what he calls a "holding environment." Heifetz provides a cogent analysis of this skill in the context of world politics (Heifetz 1994, p. 110 ff.). Paul Glen also noted the importance of this skill for managing technical professionals in his book *Leading Geeks* (Glen 2003).

When projects do get terminated, scientists will have a natural and appropriate sense of disappointment. The disappointment may become bitterness if the scientist has been working overtime and is psychologically committed to the project. When this happens (and it will), managers can help team members to effectively deal with their disappointment. Do not underestimate the value of holding formal project wrap-ups in which employee contributions are openly recognized and lauded, or having one-on-one meetings with project members to let them know that you are aware of their hard work and contributions. It is easy to neglect these small acts of recognition in the turmoil of a reorganization, merger, or change in management—but they can make the difference between the success or failure of a company adapting to the change.

Modeling behavior

Leaders who wish their teams and team members to behave in certain ways will always have more success if they themselves behave in these same ways. This is called modeling the behavior. The leader's behavior sets the norms for the team. For example, when members observe the team leader behaving in a civil manner in a tense situation, they see this as the norm and they learn to behave that way themselves. Simply telling team members that you expect them to behave in a civil manner is useless because it offers no clue for how to go about it. What you expect may seem obvious to you when you say, "I'd like to see more respect for other people's views in our meetings," but others may be genuinely baffled until you model the behavior. Moreover, if what you say is contradicted by your own behavior, its effect is nullified. We saw in the Monotech case that Andrew exhibited many of the same destructive behaviors as the group members. It therefore would not have been effective for Andrew to lecture his team on how he wanted them to behave in team meetings.

As a leader, if you find that you are repeatedly having to exhort your staff to be collaborative or "team players," chances are that you or others in leadership positions are not modeling those behaviors yourselves or are behaving contradictorily.

Focusing criticism and feedback on problems, not people

As a leader, you have the opportunity to use modeling to show the team how to attack the science without attacking one another. Where such modeling can have the greatest impact is in team meetings.

We saw in the Monotech case that the scientists dealt with their discomfort with Andrew and the project by venting their anger and frustration on one another. Yet even in circumstances without such overriding issues, scientists often address scientific matters in ways that come across as insults or attacks.

Ideally, team meetings offer opportunities for members to share the results of their work, stimulate each other's thinking, and challenge results and conclusions. Teams and their individual members vary enormously in terms of how they interact during these meetings. For example, some may choose to ignore questionable data or analyses: "I don't have the time to deal with this." Others may challenge them emotionally. For many technically minded people, sloppy work or poor experiments are annoying or worse: "You mean that you didn't even *think* about the effects of temperature [substitute 'solar radiation, high blood pressure, platform compatibility,' etc.]?" Others make personal attacks: "Don't you know *anything* about biochemistry [...low-temperature physics, infectious disease, C++]? I know undergraduates who could have figured that out."

The key to getting the most out of team meetings is to keep the focus on the problems and not on individuals. Both the leader and the participants must take responsibility for this. If the participants do not understand how damaging attacks and sarcastic aspersions can be, even a savvy leader will have difficulty keeping the discussion focused. But it is critically important that you as leader model the same behavior that you expect from the group. If you are devaluing or sarcastic, you authorize others to act that way.

Changing behavior to channel skepticism toward the problem and away from the person will enhance the likelihood that a critique or suggestion will be listened to and heard. Moving the discussion out of the personal arena increases the chances that challenging questions will be dealt with substantively rather than as personal attacks.

The following two examples show how to focus a difficult discussion on the problem, and at the same time, present a model for appropriate behavior.

Situation 1. Koshi is presenting the results of his assay. For the third time in a row, the assay has not worked. You are frustrated and beginning to think that maybe Koshi is just incompetent (so does everyone else).

- **Person centered.** "What is your problem, Koshi? I used to do this assay all the time. It's simple. Figure it out."

- **Better.** "I'm very frustrated by our lack of progress with this assay because it's really holding us back. Koshi, I need you to figure out the problem by the end of the week, all right?" This response, although it is no longer an attack on Koshi, may still show that you are angry. It also places the entire onus on Koshi. Here is another alternative:

- **Problem centered.** "We continue to lack progress in this assay, and it is really holding us back. Koshi, what help can the team provide to assist you in figuring this out?"

Situation 2. You are in a project meeting. A peer from the screening group is describing an assay that the group developed for your project. You told him last week that you believed this assay to be inappropriate.

- **Person centered.** "Shelly, why did you go ahead with that assay? I told you that I didn't want to use it and why, and you just ignored me."

- **Problem centered.** "Shelly, we discussed this assay last week and I explained why I thought that it was inappropriate. I suspect that there's a reason that you went ahead with it, and I'm interested to know what that was."

The exercises at the end of the chapter contain another example of problem-centered feedback that you can use to practice your skills.

Coping with dead air in team meetings

As we noted in Chapter 1, scientists tend to be introverted, autonomous, and independent. At times, these characteristics can make team meetings a challenge to run, since the leader expects and needs everyone's input and feedback.

Team meetings can be exhilarating or frustrating—or both. One of the most frustrating elements can be the "silent observers," who sit through the meeting and say nothing. The team leader must become skilled at persuading them to talk. A prerequisite to this is that the leader provide enough "air time" for those who are reluctant, hesitant, or shy about speaking. If you are a team leader and you dominate meetings because of your superior knowledge or discomfort with silence, you will effectively prevent some group members from speaking. One way to encourage silent observers is to direct questions to them. This shows that you respect their opinions, and you will be surprised by just how much it encourages them to contribute.

Remember also that not everyone processes new information at the same rate. Later, or even the next day, some team members may come up with an idea that relates to something that happened in a meeting. Encourage this type of out-of-group thinking as well as discussion, but be sure that the group eventually processes the new ideas.

The following case study is taken from my own career. It is an example of how self-awareness helped me to recognize that my own behavior during group meetings was contributing to a dynamic of passivity on the part of some members. Identifying my discomfort during these meetings allowed me to anticipate my behavior and choose to behave differently. My new behavior contributed to a changed group dynamic that encouraged rather than inhibited participation.

‣‣ *Case Study: The Power of Silence in Groups*

When I ran a research laboratory at St. Elizabeth's Medical Center, we held weekly lab meetings during which members presented and discussed their work. Sometimes the discussion was lively and animated, but other times it felt as though the presenter was simply telling me what he had done while the other members of the group stared off into space. At these times, I was the only one asking questions, and in place of a discussion only a two-way dialogue took place. I occasionally felt like a prosecutor dragging information from a reluctant witness. At times, this made me feel very frustrated and angry. In retrospect, I am certain that at those times I felt least secure about myself and my abilities as a leader. I felt that others were avoiding their responsibilities and forcing me to do all the work. I thought this unfair. After all, the success of our research efforts would ultimately determine continued funding from the NIH, and affect the jobs that these people held.

During this time, Suzanne, a practicing psychotherapist, suggested that I attend a weekend workshop in group dynamics. It was being held in Boston, and on a lark, I agreed.

In one of the "exercises," participants formed small groups of seven–nine people, led by a facilitator. Our task was to study and experience the interactions within our group. Period. We were sent into a room in which the chairs were arranged in a circle; when everyone was seated, the facilitator closed the door, sat down, and said nothing. This continued for what seemed like an eternity. Some of the more anxious members (I was perhaps the most anxious) started talking in an attempt to figure out what we were supposed to be doing there, as though this was some sort of a test.

I went so far as to grab a marker and list the objectives for the group on a white board. I needed to list these and reach agreement on how we would achieve them. Some joined in my frenzy of discomfort, whereas others either did nothing or periodically questioned the urgency of our need for clarity. If you read the case study of the Tea Bag Company in Chapter 1, my behavior, and the contrast between how I behaved and how some of the others behaved, will sound familiar.

This continued for about two hours that day and occurred again during two successive days. With the help of other members of the group and some frustratingly minimal observations from the facilitator, I was able to recognize and admit to the intense discomfort I felt when the group was silent. I also recognized my own need to fix the discomfort by doing something to make it disappear. I could not see that everyone—not just me—was responsible for what was happening. I could not allow the process to work itself out and for the possibility that others might have a different approach to dealing with the "task." I was too focused on my own anxiety and need for control and clarity.

During the course of the weekend, I became more comfortable with silence and the ambiguity that it contains. I became more accepting of the fact that others in the group needed to process what was going on in a different way than I. Some dealt with the discomfort by talking about their experience at that moment—a stark contrast to my approach, which was to try to change the experience into something else.

What did I learn relative to my lab? I learned to accept the times when the group is silent, to let this happen, and let others in the group take some responsibility for it. It was a revelation to realize that if I let the silence continue, which to me was excruciating, often (not always) someone would say something. When I adopted this attitude during my lab meetings, inevitably the comments that were made and the direction in which we went were quite different from where they would have gone in the past. Sometimes these comments illuminated a facet of the work that might not have otherwise emerged if I had jumped in with my own thoughts. In the parlance of the group world, I had given the others "space" to think and act.

What it took for me to change my behavior was my own experience in the model group. I needed to experience and identify my discomfort, and learn from it. I also had to try a new behavior (being quiet) and experience the consequences of that behavior. Most important, I needed to develop the ability to recognize my discomfort at the time that I was experiencing it. This recognition is what gave me the opportunity to decide what to do about it, rather than simply react to it. Finally, I needed to learn that I could live with the discomfort and that others in the group also had responsibility for the group's task. It took trust in the group for me to allow the meeting to flow, rather than try to manipulate it.

Leading productive team meetings

Because science is seen as a discipline dominated by "experts," it feels natural to senior scientists to dominate a discussion or to impose solutions to scientific problems. However, such domination can lead to a culture of passivity among more junior team members, who end up feeling that they have nothing to contribute. More important, leader-dominated discussions may constrain the scope of considered options. Daniel Feldman describes three steps to help overcome this tendency, which we paraphrase here (Feldman 1999, p. 74):

1. **Let go.** Practice letting go of the need to be seen as being right, have the only solution, and control the discussion.

2. **Solicit ideas.** Actively solicit ideas and suggestions from team members.

3. **Be open to input.** Soliciting ideas will do more harm than good if you do not give everyone's input serious consideration. Listen respectfully to concepts and suggestions from those outside of your own area of expertise or with less experience.

None of the above is to imply that actively running a meeting, or making decisions in your role as leader, is not appropriate at times. In fact, in many circumstances, such actions are essential, especially if there is a specific and clearly defined task that needs to be accomplished quickly. However, in a scientific team meeting the task is not always so well defined.

The explicit task of my lab meetings was to hear and critique the results of the lab member's work. This task involved asking questions about technique, assumptions, interpretations, and all of the details essential to the scientific process. But there was also an implicit task, which was to think creatively or at least disseminate information that would stimulate creative thought later. The leader's role in this process is to ensure that members have the opportunity to think and interact freely; often this simply requires modeling respectful behavior for the group or staying out of the way. Discussion during a lab meeting might stimulate someone to make a novel connection, generate a new idea, or ask a question that could take the group in some exciting, unexpected direction. If you are hearing a lot of comments such as, "We already tried that," "We've been over this before," or "That's a dumb idea," and very few such as, "Hey, I never thought of it that way before," examine your own responses and reactions. You may find that group members are taking their cues from you. If this is the case, experiment with the three behavioral suggestions listed above to see whether members participate more over time.

The case study above shows the way in which I learned new behaviors by participating in a seemingly artificial situation. This workshop used ad hoc groups to allow participants to experience difficult situations; role playing can achieve the same result. This is called experiential learning. The following exercises use experiential learning to help you to become more effective at managing some of the situations covered in this chapter.

REFERENCES

Avery C.M. 2001. *Teamwork Is an Individual Skill: Getting Your Work Done When Sharing Responsibility*. Berrett-Koehler Publishers, San Francisco, California.

Feldman D.A. 1999. *Emotionally Intelligent Leadership: Inspiring Others to Achieve Results*. Leadership Performance Solutions Press, Falls Church, Virginia.

Gemmill G. and Wilemon D. 1997. The hidden side of leadership in technical team management. In *The Human Side of Managing Technological Innovation: A Collection of Readings*, 1st edition (ed. R. Katz). Oxford University Press, New York.

Glen P. 2003. *Leading Geeks: How to Manage and Lead People Who Deliver Technology*. Jossey-Bass, San Francisco, California.

Hallowell E.M. 1999. *Connect: 12 Vital Ties That Open Your Heart, Lengthen Your Life, and Deepen Your Soul*. Pantheon Books, New York.

Heifetz R.A. 1994. *Leadership Without Easy Answers*, p. 104 ff. Belknap Press/Harvard University Press, Cambridge, Massachusetts.

Katzenbach J.R. and Smith D.K. 1993. *The Wisdom of Teams: Creating the High-performance Organization*. Harvard Business School Press, Boston, Massachusetts.

McAuley J., Duberley J., and Cohen L. 2000. The meaning professionals give to management and strategy. *Human Relations* **53:** 87–116.

EXERCISES AND EXPERIMENTS

1 Recognizing conflict among others

If you are aware of simmering conflicts within your group right now, list them and ask yourself whether they are impeding progress or will in the future. Pick one of the conflicts and write down three ways in which you might acknowledge or help resolve it with one or more of the people involved. Enact the conversation in your mind and notice your reactions and feelings. By remembering your responses to these imagined conversations, you will be in a better position to notice when you are avoiding the conflict in the future.

2 Recognizing conflict between yourself and others

If you are conflict averse, you may be choosing to ignore a difficult working relationship with someone in your group.

- Can you identify one of these? Determine whether it is in your own best interest, and that of your group, to continue ignoring this situation.

- List three things that you might say to acknowledge or begin discussing the problem with the other person. As above, note and remember your reactions and use these as signals to alert you to avoidance reactions in the future.

3 Using silence

- If you are a controlling manager, start by imagining what it would be like to be quiet for short periods of time in a meeting.

- Note your reaction to this imagined exercise and use this to catch yourself when you feel controlling during a meeting. Take note of what is triggering this reaction. For example, ask yourself if you are talking to reduce anxiety or control the group.

- Try being quiet for brief periods to see how you feel and people react. Does anyone who is usually silent start talking?

4 Helping others to participate

- At your next team meeting, identify those who do not contribute very much. Make it a point to ask one or more of them their opinion of some matter under discussion, whether they understand what is being said, if they have any questions, or if they have anything to add. Do this at each meeting for different people.

- After several weeks, note whether these people have become more participatory or the group is receiving input that it might not have otherwise.

5 Dealing with people who dominate meetings

The next time that you are leading a meeting in which someone is dominating the discussion, wait for a pause in the monologue to say something such as, "Karen, you're making some good points but I'm concerned that we're not hearing everyone's views. I'd like to hear what other people have to say. Jim, what are your thoughts?" Does this approach brings others into the discussion? You may need to have a conversation with Karen in which you focus on the problem, as illustrated in the next exercise.

6 Attack the problem, not the person

The following is a hypothetical conflict that requires attacking the problem, not the person. For the exercise, we suggest that you start by attacking the person, so that you can get some idea of how this would sound. Then attack the problem as described in this chapter.

> You are explaining a complicated concept to your technician so that he can better understand his project. He does not seem to be listening. He is not looking at you and he is fiddling with his Palm Pilot. You have noticed that since he has been hired, he does not seem to pay attention to what you say, and goes ahead and does what he thinks is best.

What would you say to address this conflict that attacks the person? What would you say that attacks the problem? If you wish, write down what you would say or do in each case. Suggested answers follow below.

- **Ineffective responses.** Use avoidance; do not say or do anything. Do not speak to the technician; let him do what he wants.

- **Attacking the person.** "You're not listening to what I'm saying. In fact, you never seem to listen to what I have to say. I feel like I'm talking to the wall. If you don't shape up, you will be replaced."

- **Attack the problem.** "We don't seem to be communicating well. I notice that when I talk to you, you don't really pay attention. Is that true? [He may tell you that he *is* listening.] It would help if you gave me some indication that you are listening to me. Is there anything I can do to be clearer? [He tells you that he is better at seeing things than listening to words.] Let's go to the white board and I'll describe it to you there. Then we can go back to the lab and I'll show you."

    Write down a problem that you have with someone in your team or group.

- What are one or more things that you could say to attack the person?

- Write down your likely responses or reactions. Create a dialogue that allows you to experience this hypothetical conversation. Using self-awareness, try to capture your feelings during this dialogue. You may find that you were angry, and that your attack was a reaction to that feeling. **Practice noticing when you feel anger or frustration as a result of someone's behavior. Once you notice your anger, you can consciously redirect it to the problem.**

- What are one or more things that you could say to focus on the problem?

- Write down your likely response. Create a dialogue that allows you to imagine this hypothetical conversation.

Were you able to find a way that clearly communicated the problem using the second approach? If not, reread the examples in this chapter to find other ways of presenting or explaining the problem.

Additional exercises for attacking the problem, not the person, are given in Chapter 5.

CHAPTER **5**

A Delicate Art:
Manage Your Boss

Scientists who become managers and leaders of technical organizations are cut from the same cloth as those they manage. Like everyone else, their blind spots and limited self-awareness can prevent them from picking up important interpersonal cues and contribute to conflict aversion. Nonscientists who manage or lead scientists may have strong interpersonal skills, but can be at a disadvantage if they do not understand scientists or how science is done. This chapter shows how you as a scientist can navigate the choppy waters of interacting with your boss, whether she is a scientist or not. It also shows how you can turn interactions that start out as confrontations into productive problem-solving sessions. Finally, you learn how to understand your boss's problems, deadlines, and goals; disagree with him or her; and respond to criticism.

▸ **You and authority: Is there a pattern?**
▸ **Hidden boss traps**
▸ **Dealing with an angry boss**
▸ **The three tools**
  Manage the boss's anger or frustration
  *Identify and address the interests that lie
    beneath the issue at hand*
  *Be cognizant of your role and
    responsibility*
▸ **If your boss is not a scientist**
▸ **Handling feedback**
▸ **References**
▸ **Exercises and experiments**
  *1. Pattern recognition: Your interactions with
    authority*
  2. Misattribution of your boss's behavior
  3. Working with an angry boss
  4. Disagreeing with your boss
  5. Understanding your boss's needs
  6. Criticizing your boss
  7. Think before you speak
  8. Your boss's perspective
  9. Boss traps

YOU AND AUTHORITY: IS THERE A PATTERN?

Self-awareness can enable you to decide whether feeling insulted or hurt is caused or magnified by your own emotional filters. The benefit lies in ensuring that your actions and reactions are appropriate to the circumstance. Nowhere is this benefit more important than in interactions with your boss.

People in authority can elicit strong reactions in us. But in many cases these reactions have more to do with associations to a parent or other authority figure from childhood than with the person himself. These reactions may lead to behaviors that get in the way of our working relationships with those in authority. Becoming aware of your own reactions to authority figures is the first step in deciding whether your interactions are influenced more by the past than the present.

For the great majority of people, preexisting reactions to authority figures originate with a dominant parental authority figure, often, but not always, a father. How might your reactions to your boss be related to relationships with childhood authority figures? Let us look at some hypothetical examples.

1. If you were fearful of a childhood authority figure, you may find yourself avoiding or fearing your boss. Perhaps what your boss does or says does not seem to cause others concern, but feels threatening and anxiety provoking to you. These kinds of reactions may lead you to avoid your boss, or to overreact to comments or actions that others find benign.

2. If childhood authority figures did not command or earn your respect, you may find it hard to respect your boss. You may be skeptical of what she says, poking fun at her behind her back or ignoring her directives. Because you have no conscious intention to disrespect her, you may be surprised when she confronts you with your behavior.

3. If authority figures were either absent or paid little or no attention to you as a child, you may find yourself acting as though you do not have a boss, even though you really do. The following brief case study illustrates this type of behavior and its consequences.

▸▸ *Case Study: Ignoring Your Boss*

Ulla was a senior group director at the Nanotech research center. She reported to Yoko, the vice president for research. Yoko hired Ulla because she was both highly productive and independent. They worked well together until Yoko took a particular interest in the largest of Ulla's projects that was beginning to become very costly. Yoko was careful not to interfere with Ulla's leadership of the project, but several times during the last three months, she reviewed Ulla's expenditures with her and recommended alternative approaches to cut costs. Ulla listened politely to her suggestions, but did not implement any of them. As the year progressed and budget time approached, Yoko learned indirectly that Ulla was planning an expansion of the project, doubling its budget.

Yoko told Ulla that she had one month to improve her communications with her and show that she could be responsive to Yoko's directives and suggestions. She suggested that Ulla meet with the head of human resources to develop a plan to achieve this.

Ulla was shocked by this turn of events and completely unaware that she was behaving as though Yoko did not exist. However, once she discussed her behavior with the head of human resources she was reminded of a similar problem that she had had with her Ph.D. adviser in graduate school. With the help of human resources, Ulla came up with a plan to improve her working relationship with Yoko. Ulla proposed that she meet with Yoko weekly and that she copy her on any important correspondence related to project expenditures and budget. Ulla felt that routine contact with Yoko would help her to maintain an awareness of their relationship and prevent her from behaving as though she did not report to Yoko. Ulla had to work at reminding herself to touch base with Yoko on a regular basis, but with time Yoko saw an improvement in Ulla's responsiveness and the crisis passed.

Because I know Ulla well, I can tell the reader that as a child she had very little contact with her father, who was the central authority figure in her family. He was regularly absent from the home, and when he was there, he was often the butt of her mother's disrespectful comments. It is no wonder that Ulla grew up believing that authority could be safely ignored and she needed to fend for herself if she wanted anything. It was not that authority figures made her angry or threatened her, it was simply that she had no use for them.

Not all of us have, nor do we necessarily need, the kind of insight into our past relationships with authority figures that Ulla had. All we need is an ability to detect patterns in our current interactions with authority figures. When Ulla remembered that she previously had a similar problem with her thesis adviser, she concluded that her disregard for authority figures was a recurring pattern to which she needed to pay close attention in the future.

Interactions like those between Ulla and Yoko are never unidirectional. Yoko also has some responsibility for the events. She encouraged Ulla's independence, perhaps to lighten her own supervisory burden. As long as things were going well, Ulla's independence was not an issue for Yoko. This independence felt very comfortable to Ulla, who, as we have seen, would just as soon have ignored Yoko altogether. Only when budgetary matters caused Ulla's and Yoko's interests to diverge did the problem emerge.

This case study highlights the importance of being alert to behavior patterns from past relationships with authority figures for avoiding those same behaviors in the present. As you become adept at this, you will detect patterns in your relationships with authority figures over time and in different settings—with teachers, advisers, mentors, project managers, CEOs, etc. Moreover, you will be able to foresee or anticipate problems before they arise, proactively managing yourself and your relationship with your boss.

HIDDEN BOSS TRAPS

Some problematic behaviors such as being demeaning, overly demanding, inconsiderate, and manipulative are relatively easy to spot in bosses. But other behaviors that seem harmless on the surface can produce consequences every bit as negative. If you are engrossed in your work and have difficulty seeing beneath overt behavior patterns, you may find yourself manipulated, abused, or misled by your boss. It is in your best interest to learn to recognize questionable behaviors before they get the best of you. Here are some patterns to look for.

- **Boundary issues.** Your boss talks to you about her personal problems. This makes you feel special. But it also leaves you with a vague feeling of discomfort that you brush aside. Pay attention to that feeling; it is telling you something important. Developing a personal relationship of this type with your boss will almost certainly compromise your professional relationship with her.

- **Playing favorites.** Your boss plays favorites and you are the current one. You are put in charge of projects, get to make important presentations, and go to scientific meetings in exotic places. You feel good about this, but others are not treated as well, and you feel uncomfortable about that. You brush aside those feelings and focus on your work. But it is best to pay attention to those feelings—it may be that soon someone else will be the favorite and you will be disappointed or angry.

- **Selective communication.** A boss who plays favorites may also disseminate important information to selected members of the organization, making everyone else's job harder and progress slower. If you are on the select list, you feel fine. But if you are on the left-out list, you may not get the information or communications needed to do your job. The result is detrimental to you and the company.

- **The boss as booster.** Because she likes you, your boss promotes you to a position for which you are really not qualified. At first you feel uneasy, but you quickly bury those feelings as you struggle to keep your head above water in your new job. In time you feel that you have been set up to fail because you lack the experience and skills to succeed. Next time, pay attention to your feelings right away, and question the advisability of such a promotion as well as the motives behind it.

If you are disposed to look only at the manifest behavior of your boss you may be blind to the negative consequences of such behaviors. If you attune yourself to your feelings, doubts, and misgivings, you increase the likelihood that your work and career will not be subverted by your boss's errors in judgment. If he is powerful and charismatic or his questionable behaviors mesh with your own needs (for approval, recognition, acceptance, etc.), you will have a hard time recognizing the downside of those behaviors. As you become more aware of your emotional needs, your ability to make dispassionate assessments of your boss's behavior and its appropriateness will improve.

After recognizing these behaviors, take steps to limit their impact on you and your team, project, and organization. For example, in the above cases, you can say the following:

- "Fred, I'm not sure that I'm the right person for this errand. I don't feel comfortable doing it."

- "Fred, I appreciate your confidence. I'd like to share this information with the rest of the team," or "Fred, getting information in this way puts me in an awkward position with the rest of the group. I know that they would really appreciate periodic meetings in which you could update them."

- "I'm really honored by your confidence. I'd like to try this unofficially for a month so we can both decide whether this promotion makes sense."

DEALING WITH AN ANGRY BOSS

The previous sections illustrate some of the ways that limited self-awareness can lead to inappropriate behaviors that result in misunderstandings or clashes with bosses. Moreover, if your boss is a scientist, he may be suffering from the same deficits in self-awareness and interpersonal savvy as you. Limitations like these, that occur in both employee and boss, can lead to escalating problems and misunderstandings, possibly resulting in overt clashes. If this happens, you must defuse the hostility and work toward reestablishing the relationship.

When dealing with your boss, remember that your goal is a productive interaction, not an argument or confrontation. To make this happen, remember to manage your own reactions. In addition, you must change a confrontational atmosphere into one in which the two of you become allies who seek to solve common problems. These problems are defined by the underlying interests of you and your boss.

The following case study illustrates the application of this approach. It involves an interaction between a graduate student and her mentor (it has direct relevance to the topics on mentor-student interactions covered in Chapter 7). Although the mentor-student interaction has some unique characteristics, it is also has elements of an employer-employee interaction. The behavior of the employer/mentor in the following may seem extreme, but the case, as related to me by a close colleague, is based on actual events. It shows how a junior scientist might have dealt with an infuriating situation.

►► Case Study: *When Your Future Is on the Line*

Barbara was a graduate student whom I met during a physics class that I was teaching. Around the middle of February of her third year in graduate school, she e-mailed me to ask if I would discuss a problem that she was having. I knew that Barbara was a member of Professor Aster's laboratory, where she was working on a Ph.D. thesis project. She had been there for about 18 months. My assumption was that this meeting was to be a scientific consultation, since I knew that Aster's scientific interests to some extent overlapped with my own.

The problem turned out to be only peripherally scientific. Barbara told me that she was having a hard time with Professor Aster because she was not satisfied with the amount of time Barbara was spending working on her project. Aster had two other graduate students and two postdoctoral scientists in her lab and it seemed to Barbara that they all worked 12–18 hours per day, six or seven days a week. Barbara was married and wanted to have a life outside of the lab. She felt that she put in a reasonable amount of time in the lab, but this apparently was not enough for Aster.

The previous week, Aster had called Barbara into her office and in the course of a rather rambling conversation, suggested that Barbara was perhaps not cut out for a life in scientific research. Aster said that some women were happier staying home and having babies, and that maybe this was where Barbara belonged. Barbara was shaken up and distraught by this interaction. As she relayed the story, I could see her becoming visibly upset and overcome with emotion.

I asked Barbara how she had responded to Aster's outrageous comments, and she said that she was shocked and hurt. She was so hurt that she became immobilized; her feelings overpowered her thoughts and she could say nothing. After some moments of silence, Aster announced that she had another meeting to go to and the interview ended abruptly.

I can still recall the feeling of ice in my veins when Barbara relayed this story to me. Barbara did not know what to do. She had invested 18 months on this project and did not intend to walk away from it. She suspected that Aster wanted her out of the lab and was unsure that she could do anything about it. Barbara felt guilty for not working as many hours as the others but was not about to sacrifice her personal life or her relationship with her husband for a few more hours in the lab.

I was seething and wanted to tell Aster that her behavior was outrageous and possibly unethical, and that I was going to bring her before a faculty committee for censure. However, as soon as I became aware that I was fuming inside, all my self-awareness alarms went off. I knew from experience that taking action while feeling this way would be a bad idea, both for me, and more importantly, for Barbara.

I asked myself what was of paramount importance here. Was it to punish Aster for her inappropriate remarks or help Barbara get through this situation and obtain her degree? I concluded that no matter how much I and Barbara longed to make Aster accountable for her behavior, Barbara's overriding interest now was to complete her project and get her degree in the shortest time possible.

I suggested to Barbara that she needed to weather this storm. I told her that Aster was probably under stress and that Barbara should not take her comments literally. Since I was on her thesis committee, I knew that her work was moving forward, even if it was not at the rate that Aster might have wanted. I told Barbara to initiate a meeting with Aster sometime during the following week for a follow-up discussion and ask Aster to meet with her more regularly to give feedback. I suggested that Barbara make it clear to Aster that she was in the project for the long haul and had every intention of completing her work there. Barbara did this and continued to work hard on her project. She completed a first draft of her thesis 18 months later, and six months after that, she left Aster's lab for a postdoctoral position.

What could Barbara have done differently in this situation? Although she could have made a major issue of Aster's behavior, that would not have been in her best interests. Had Barbara defined her underlying interests as described in Chapter 3, she would have concluded that her chief interest was to complete her Ph.D. work in two more years. It was almost a certainty that seeking vengeance on Aster would have initiated a chain of events that would have had a major impact on this goal. Further, Barbara was not about to quit the lab, since it would result in an unacceptable delay in completing her Ph.D. work. Moreover, Aster's underlying interest of a productive lab would not have been satisfied if Barbara left the lab prematurely. After all, Barbara was doing her work and Aster had invested a significant amount of time and resources in her project. Both of them tacitly understood that each would lose if the relationship deteriorated.

In this instance, my colleague's support and suggestions helped Barbara to negotiate a successful outcome with Aster. Barbara had sufficient self-awareness to know that she needed help in this situation, and sufficient observational skills to identify the right person to help her. It is possible that if Barbara had done nothing in response to the meeting with Aster the situation would have turned out fine. It is also possible that if Barbara had not come to my colleague for advice, the situation would have deteriorated and she would have left Aster's lab, setting her education back significantly. We will never know.

But we do know that if Barbara could have done or said something during her meeting with Aster to reduce the chance of a disastrous outcome, she should have. In the following sections, we provide three approaches or tools for dealing with an angry or frustrated boss. Of course, the tools can be applied to anyone with whom you are interacting. We present them here because most of us have greater difficulty dealing with an angry boss or authority figure than with an angry peer or subordinate. If you can get it right with your boss, you are likely good to go with almost anyone else.

THE THREE TOOLS

For Barbara to use the following tools, she first needs to recognize and manage the strong reactions that Aster's comments have engendered within her. Using the techniques described in Chapter 2, Barbara might have been able to recognize what she was experiencing and feeling, and anticipate from past experience that her reaction would be to withdraw and be silent. If she can get to that point, she will be better able to apply the approaches listed below. First, we briefly describe each technique and then illustrate them in more detail.

Manage the boss's anger or frustration

A productive outcome from a difficult interaction with your boss requires care to avoid exacerbating the situation. We offer three tactics to help you to manage your boss's anger and focus the discussion on the underlying issues.

Identify and address the interests that lie beneath the issue at hand

As we saw in Chapter 3, this is a central tenet of negotiation and an important element in almost any difficult interaction. By learning to look beneath what may seem like impossible or inappropriate demands, you are in a better position to come up with creative solutions to your boss's problems.

Be cognizant of your role and responsibility

Ask yourself, "What is my role and responsibility in this situation?" Meeting your responsibilities is one of your underlying interests and asking yourself this question may forestall a knee-jerk reaction. I think of the sign that a friend used to display in his office: "Put brain in gear before putting mouth in motion."

In the preceding case the lack of resolution during the meeting was due to Aster's inability to provide useful feedback and Barbara's inability to turn a confrontation into a problem-solving session. What might have happened if Barbara had followed the three guidelines below?

Tactic 1. Manage the boss's anger or frustration

To arrive at a resolution of the problems presented by Aster, Barbara needs to shift the tone of the discussion from anger and accusation to joint problem-solving. She must stop feeling like a victim and show Aster that she has heard her and is sympathetic to her concerns. The following three guidelines, taken from *The Feeling Good Handbook* by David Burns (1990), work wonders in this and similar situations.

Agree. It is hard to be angry with someone who is agreeing with you. For Barbara to have a productive conversation with Aster, she must defuse Aster's anger from the outset. Finding something with which to agree on with Aster is a great way to do this. It lets Aster know that Barbara has heard what she has said. Starting the conversation in this way will make Aster more receptive to what Barbara has to say. Despite Aster's rather primitive attack, Barbara can still find something with which to identify and agree. She might say, "It's true that I don't work as many hours as Linda or Barry," "I agree that Barry and Linda seem to work to all hours of the night," or even, "This has certainly been a slow month for data collection in the lab." Barbara has admitted to no inadequacy, but has begun to establish a rapport with Aster. By showing that she shares Aster's concern about productivity, Barbara can start to defuse Aster's anger. Once anger is defused, Barbara can build a bridge for communication by empathizing with Aster.

Empathize. Barbara can indicate that she acknowledges Aster's frustration and can understand why she feels that way. Barbara could say that she is aware of the pressures and deadlines to which Aster is subject. Perhaps she has noticed Aster's concern over the past several weeks and understands why she is concerned. She could say that she knows that this has been a slow month for either her or the lab as a whole and understands why this makes Aster uneasy. Any of these statements sends a clear message to Aster that Barbara cares about what Aster is saying. Once she gets to this point, Barbara is ready for the next step in the process: assurance.

Assure. Barbara can assure Aster that she is committed to doing her part (without saying that she will work 20 hours per day) and understands the need for greater productivity. If these statements are genuine, they will strengthen the bridge that Barbara is trying to build.

Inquiry. Barbara is now ready to move to problem-solving, the final stage of the process. She can start by asking problem-solving questions, e.g., what Aster would like to see as the outcome of the conversation, what she should be focusing on during the next month, or for ways other than working 20 hours per day to show Aster her level of commitment. Of course, if Aster really does believe that everyone in her lab should be working 20 hours per day, the outcome may be unsatisfactory. The inquiry phase is especially important because it can help uncover the underlying causes of Aster's concerns.

Remain focused on your task when dealing with an angry or hostile person

Your task is to get your work done, not to act insulted. However, it is difficult to work in the presence of a hostile, argumentative person. Therefore,

- **Agree** to defuse anger
- **Empathize**. Start a dialogue
- **Inquire or assure.** Show interest and be positive
- **But** do not give in to pressure

Tactic 2. Identify underlying interests

As we saw in Chapter 3, identifying underlying interests can be the most important element in a negotiation. Aster demanded long hours to satisfy some underlying interest. Through inquiry, Barbara may learn that Aster is under pressure to produce data during the next three months to ensure the renewal of an important Department of Defense grant. If this is the case, Barbara might be willing to agree to work seven days per week for the next six–eight weeks, with the understanding that this is a temporary arrangement.

It may be that Aster is feeling anxious because of a series of setbacks, and her underlying need is for reassurance that everyone is fully committed to the project. In this case, Barbara's expression of understanding and support may provide such reassurance.

It may be that other postdocs in the lab are resentful that they work longer hours than Barbara and have complained to Aster. In this case, Aster's underlying interest is to defuse tension in the lab. Barbara can suggest ways that she could interact with the other postdocs that would show them her commitment to the project.

Tactic 3. Be cognizant of your role and responsibility

Barbara has at least two roles in this case. One is that of a trainee and the other is a member of a team working on a project. In her first role, she must keep her training program on track so that she can finish within two years. This requires maintaining her relationship with Aster. In her second role, she is responsible for helping advance a group project. Barbara must let Aster know that she understands her responsibility to the group effort and is willing to pitch in (within reason) to help advance it.

Reminding yourself of your role and responsibility in a difficult situation is often an effective way to gain some perspective on an emotion-laden interaction. It forces you to step back from the fray and look objectively at your responsibilities. This can be a valuable tool for interrupting knee-jerk reactions until you have had enough time to think through your options. Ask yourself

- What is my responsibility to the team, or organization?

- Am I acting in accordance with that responsibility?

- Am I contributing to the work of the group or team?

In summary, if all Barbara can think of in response to Aster's confrontation is how she can spend more time in the lab to generate more data, she may have missed the issues that underlie Aster's outburst. If Aster's frustration with Barbara's working hours reflected her belief that Barbara was not committed to her work and was not interested in helping the lab to maintain its funding, Barbara needed to do more than just work longer hours. Judicious use of the "inquiry" tactic during the discussion with Aster could help Barbara look beneath the surface and seek the underlying reasons for Aster's frustration.

IF YOUR BOSS IS NOT A SCIENTIST

One of the hardest challenges facing a scientist can be working for someone who is not herself a scientist. Scientists tend to be concrete, linear thinkers. They are cautious in their conclusions and may be loath to make the kinds of projections, speculations, and leaps of faith that come naturally to those with more of an entrepreneurial bent. Executives with business or entrepreneurial backgrounds may make connections that seem grossly speculative or even irresponsible to scientists. They may be willing to take risks to take advantage of opportunities. These executives may become easily frustrated by scientists who strive for certainty and thoroughness. Conversely, scientists may act in a condescending manner toward an executive who controls their fate, but who they see as being scientifically naïve.

Scientists must understand that nonscientist CEOs may have only a vague understanding of how research is done. Some are skeptical when a scientist implies that she cannot guarantee that a discovery will be made or a promising therapy can be completed in 12 months. In a sense, the scientist and the executive are speaking different languages. As a result, a hidden dialogue takes place when they interact. The following is an illustrative case study.

▸▸ *Case Study: Speaking Different Languages*

Pharmex is a successful biotechnology company employing more than 500 people. It makes extensive use of metrics to monitor productivity in its drug discovery division. Last year was a difficult one for the company. Two promising drugs that were either in or about to enter the clinic had been withdrawn because of safety problems. The company's stock had dropped, and industry analysts were skeptical that Pharmex could improve its profitability for at least two years given its existing pipeline of drug candidates. Alex, the CEO, wants to show the board that the company is not in as bad a shape as it seems, and plans to give an upbeat presentation about some of the more promising project possibilities. He has asked Luisa, the head of his cardiovascular research division, to meet with him to discuss presenting the progress on a cardiac arrhythmia drug that Pharmex is developing. The overt dialogue follows.

Alex: "Luisa, we have a board meeting coming up in two weeks and I want you to give them an update on the AR 287 project. We've put this off twice and now we need to bite the bullet. Put together some time lines that predict when you believe it will be ready for its first trial in man."

Luisa: "Well, that will depend on a lot of factors. We're still trying to find a good biomarker for activity in man. That has been a slow process because our expression profiling group spent six months on something that turned out to be bogus and now we're back to square one. I really don't have anything concrete to tell the board."

Alex: "You need to do better than that, Luisa. Can you just make some projections for us? I mean, you must know how much work you have to invest to come up with a biomarker, whatever that is."

Luisa: "Alex, it is simply not possible to project when discoveries will be made. You just don't understand research."

Alex: "Well, I understand this: Either you figure out how to give us a useful update and time line for the rest of this project, or I will find someone else to do it."

Luisa's responses to Alex may have been honest, but they did not achieve much more than to create additional anxiety for him and lead to an ultimatum for her. Let us imagine what was going through each of their minds as they had this discussion. Here is the covert dialogue:

Alex: "Here we go again; this is like pulling teeth. How am I supposed to develop a budget and a time line for the board if these scientists never commit to anything? What can I do to get a straight answer? What do they do in those labs all day, play computer games?"

Luisa: "Here we go again. Whenever I talk to this guy, it feels like a root canal without anesthesia. He just doesn't understand that it is not possible to guarantee anything in research. All he wants is projections and time lines. Am I supposed to make up numbers? I wish that he would just back off and let us get on with our work."

Because Luisa is focused on the science, all she hears in Alex's request is a demand for results. Although management needs to see real results, Luisa feels that she cannot tell them what she does not know. By limiting her focus to the scientific impediments to meeting Alex's demands, she has lost sight of what lies beneath these demands. Only by discovering Alex's underlying motivation, interest, and needs will she be able to help him and fulfill her role and responsibility. Let us see how Luisa might use some of the approaches discussed above in this situation.

Manage Alex's frustration or anger (agree/empathize/inquire)

- Luisa can **agree** with the need to present concrete information to the board. In this way, she confirms the need to present something but she does not say that she has such information. She can also agree that it is unfortunate that they have not been able to provide concrete information to the board on this project in the recent past.

- Luisa can **empathize** with Alex's dilemma. This will go a long way in reassuring Alex that they have the same objectives. She could say, "I know that you are under the gun here and that the board would be thrilled to see positive results and projections on this project. I want to help and will do whatever I can to pull together an accurate and realistic picture of our progress. Although you know that I cannot pull a rabbit out of a hat, I can be upbeat and positive since that is how I feel about this project." Hearing this from a scientist will make most CEOs think that they have died and gone to heaven. Luisa will have made a friend, while being honest and frank about what she can do.

- Finally, Luisa can **inquire** about the specific type of information that Alex would like to hear in her presentation. Further inquiry will also help her to understand the issues beneath his request. Does he simply need data on AR 287, or is he trying to meet other underlying needs? Whereas it may sound as if he is asking for something that Luisa cannot deliver, he may actually be asking for something much simpler—an indication that she understands his problems and that she is willing to work with him to help address them.

Identify Alex's underlying interests or needs

The following lists the possible interests that underlie Alex's request:

- Demonstrate progress on this specific project to the board.

- Get reassurance from Luisa that the project is moving forward.

- Get Luisa to commit publicly to a specific list of deliverables and deadlines for the project.

- Showcase his ability to motivate and inspire research and development in the company.

- Use the board meeting and Luisa's presentation to help him decide between continued funding of this project versus others.

Luisa can get a good idea of the interests that underlie the CEO's request by asking questions. On what would Alex specifically like her to focus? What should be the underlying theme of her message to the board? Should she show data or just review progress and projections? How much technical detail should she include? Should she discuss the cardiovascular group's alternative to this project if it fails to progress?

Of course, you can access information other than by asking for it outright. If Luisa is attentive to the activities of the company, she may already know or suspect what (beyond a clearcut progress report) Alex is trying to achieve by scheduling her presentation. She may also suspect that it is just a straightforward presentation, but if that were the case, Alex might not have been so annoyed at Luisa's obstinancy.

By addressing Alex's underlying needs (as opposed to his demand for results), Luisa can deliver something other than or in addition to results (that she believes she cannot deliver). If in response to some of her inquiries the CEO says, "Look Luisa, you need to give us some indication that this project is moving forward and we have alternative strategies in place if it continues to stall," Luisa can reasonably conclude that Alex is looking for reassurance. This does not mean that she should dissemble about results, but that in this circumstance, Alex's demand for results may mean something in addition to data. Perhaps it signals a review of the project plan, discussion of steps taken to ensure that the project is keeping to its time line, review of contingency plans, and her assessment of likely outcomes. These all add up to reassurance that all is under control.

We have two final points on this topic. First, remember that both nonscientists and scientists may use the same words to mean different things. For example, when a senior executive (especially one without a science background) asks for an update on a project, she almost never means, "Show me your latest data." She is looking for the big picture. If you present the big picture succinctly, accurately, and honestly, you will be the one this executive comes to the next time she needs information. If you open your lab notebook and start flashing micrographs of a rat's liver or pictures of polymerase chain reaction products, do not be surprised if you are never asked again.

Second, executives with little or no science background may have a different conception of time than do scientists. Dubinskas (1988) explains that managers with a business background tend to view time in a linear manner, punctuated by well-defined targets and milestones. Scientists, on the other hand, view time as having a flexibility that reflects the reality of scientific research: You cannot schedule discoveries. These viewpoints need not be incompatible, but both parties must understand that words or phrases such as, "as soon as possible," "quickly," and "fast track" may have different meanings for each of them. If not, confusion and disappointment may result.

HANDLING FEEDBACK

At some time or another most of us receive feedback from someone in authority, either in a formal performance review or via an informal comment. Such feedback is one of the most important sources of information you can get about your behavior, effectiveness, and work performance. It is unfortunate but true that most people tend to think of offering feedback only when something is wrong or your performance is not up to expectations. For such situations, knowing and anticipating how you react to negative feedback is the best preparation. In addition, here are some responses to feedback that you can use to increase your chances of taking the best advantage it.

• Repeat the feedback in your own words and ask if you have it right. Often we hear something other than what was actually said, especially when it has an emotional impact on us. So if your boss says, "Jim, your presentation to the investors who were here last week wasn't at all what I expected. It was over everyone's head and it needs a lot more work." You might respond with,"So you're saying that you expected a less technical presentation and you'd like me to rework the one I gave to make it more suitable for a lay audience, right?" You might be correct or you may get the response, "No, I don't mean that I want you to pitch it for a lay audience. I just think that you need to cut down on the number of equations you show." That response would make you thankful that you clarified his feedback instead of just going ahead and acting on your assumptions.

• Before responding to feedback, especially negative feedback or criticism, find something with which to agree. We discussed the rationale for this earlier in the chapter, and the same applies here. Your goal is to bring a positive and collaborative tone to the discussion. So Jim might have said, "Well, I did notice that a few people were seemingly baffled while I was talking," or "I guess my scientist side sometimes gets the better of my communicator side." After this, you can inquire about the feedback and identify underlying interests: "Did one of the participants say something to you?" or "Do you think that they missed the point of what I had to say because I bogged them down in the equations?"

  Next, using nonconfrontational problem-centered words, identify aspects of the feedback with which you might not agree. For example, "I was actually following the guidelines that your assistant sent out before the meeting. I may have misunderstood them, but if we look back at them, I think you'll see that they advise citing detailed scientific support for our conclusions, including equations."

- Identify ways to solve the problem together. What do you need from your boss to help you improve or change? It may be that you need increased feedback or more frequent meetings. **Keep the focus on what you or your boss needs to do to improve the situation in the future, rather than on what is wrong with the situation now.** Jim might suggest, "Perhaps if I had a bit more regular feedback from you, I would do a better job of knowing how you want these presentations structured. What if we meet once a week during the next month or so while we're in a fund-raising mode?"

  Finally, ask your boss to name what you are doing right so that you have a concrete idea of what he wants. In Jim's case he might have asked, "Were any of the presentations I gave last month closer to what you had in mind?" or "The equations aside, was the rest of the presentation on target? What were the strongest parts, so that I can better gauge what it is you wish to see?"

The exercises that follow help you to gather data about patterns of behavior that you may exhibit when interacting with authority figures. They also include experiments for exploring new behavior patterns and determining whether these can make your interactions with your boss more productive.

REFERENCES

Burns D.D. 1990. *The Feeling Good Handbook.* Plume/Penguin, New York.
Dubinskas F.A. 1988. *Making Time: Ethnographies of High-technology Organizations.* Temple University Press, Philadelphia, Pennsylvania.

EXERCISES AND EXPERIMENTS

1 Pattern recognition: Your interactions with authority

Write down the names of three authority figures from your professional career. These could include your current boss/supervisor/manager and any from the past. Next to each name write the first four feelings that this person evokes in you. Then answer the following questions:

- Do your feelings toward these different authority figures have anything in common?

- Do these common features suggest that you might be reacting to feelings (perhaps the ones that you have identified as common to several authority figures above, or others) that are evoked by your boss, but that originate from your past, i.e., are they only loosely related to your boss?

- If so, do these reactions affect your work or the effectiveness of your interactions with your boss?

2 Misattribution of your boss's behavior

Often, we assume that something a boss says that we find upsetting or objectionable was specifically directed at us, or in response to something we did. Sometimes just enumerating alternative explanations can help us to see that these are just as, if not more, likely than the "self-centered" explanation.

List three things that your boss has said or done that angered or threatened you. For each incident, list three explanations that have nothing to do with you and three reasons that have everything to do with you. Evaluate the likelihood of each set. Often, this exercise will enable you to make a more accurate assessment of the situation and identify inappropriate or exaggerated reactions on your part.

3 Working with an angry boss

The next time your boss says something that seems angry or threatening,

- Say one thing that conveys an understanding of your boss's issue.

- Find one thing with which to agree.

- Ask one clarifying question.

- Check whether these techniques change the dynamic of the interaction.

4 Disagreeing with your boss

- The next time your boss says something with which you disagree, find some element in what was said with which you do agree.

- Start a conversation by talking about that element.

- Transition to what you do not agree with in a way that focuses on the impact of the decision, statement, etc. on the project or company.

- Assess whether this approach helps you to convey an alternative view in a way that is collaborative rather than argumentative.

5 Understanding your boss's needs

- Think of something that you know that your boss needs from you, but has not yet asked for, and do it.

- If you are focused on your work or on technical matters, it is easy to lose sight of the fact that your boss needs information, viewpoints, and support from you. Forcing yourself to think like your boss or put yourself into his or her shoes is one step closer to creating a more productive relationship. Make a list of three things that your boss might need from you, especially what he might not know he needs, and give him one of them before he asks.

- Take note of whether your new behavior changes the tenor of future interactions with your boss.

6 Criticizing your boss

- List one or two comments or behaviors that you have recently said or done that belittle or denigrate your boss.

- What was the stimulus for these actions? What were you feeling when they happened?

- Review your authority behavior pattern (earlier in the chapter) and ask yourself whether your behavior was motivated in part by feelings unrelated to your boss.

7 Think before you speak

- Catch yourself the next time you feel anger, hostility, or animosity toward your boss. Instead of acting on your feelings, remember them.

- Later, review what you would have done or said if you had expressed the anger, hostility, or animosity, and the possible consequences.

- If, as is likely, you decided that you were better off not acting on your feelings, remember this the next time you experience these feelings.

8 Your boss's perspective

It is easy to resent what your boss says or does especially if it negatively impacts you. However, sometimes resentment is the result of ignorance of the pressures and constraints under which your boss works.

- List the external and internal pressures and constraints to which your boss is subject.

- Pick one decision that your boss has made with which you disagreed or that had a negative impact on you. Take your boss's perspective and write down as many reasons as you can that might have influenced his or her decision.

- The next time your boss makes a decision with which you disagree, remember the results of this exercise, and list the possible underlying circumstances or explanations.

9 Boss traps

- List what your boss has said or done that has made you feel even mildly uncomfortable, and that was outside the scope of good management practice and behavior. Include comments or behavior that did not seem overtly inappropriate, but that just left you with an uneasy feeling.

- Can you discern recurring behaviors in this list?

- Use this list to identify future situations in which you may wish to exercise caution in your interactions with your boss.

CHAPTER 6

Win/Win with Peers: Make Allies, Not Enemies

arly in my career, I was one of a three-person team working on an exciting research project. Midway through the project, we argued about who would be first author on our scientific paper. We went to our lab director, who said, "Work it out among yourselves. I have faith that you can do it." But he was wrong: We could not. The publication of the paper caused great consternation within the lab. The team disbanded and the project quickly lost momentum. We all lost.

Some of the most difficult interactions that scientists face involve peers and colleagues. Unlike interactions with employees or superiors where one party can always make the final call, peer interactions take place on a level playing field. It is here that negotiation skills are especially important because it may not be possible for any of the parties involved to make an autonomous decision. Inability to resolve a dispute may require intervention by a higher authority, such as a director or administrator. When problems escalate to this level, all semblance of civility is frequently lost and the consequences can be disastrous, not only for the individuals, but for entire laboratories, departments, and companies.

- ▸ Peer problems in the science workplace
- ▸ Key elements of peer conflict avoidance
- ▸ Tools for dealing with peer conflict
  Recognize the warning signs
  Avoid accusations by addressing the problem in terms of its effect on you
  Address the specific behavior without devaluing or belittling the person
  Defuse anger and hostility
- ▸ How to disagree about science without arguing
- ▸ Pay attention to your own body language
- ▸ Pay attention to the other person's body language
- ▸ Consequences of feeling inferior/superior to peers
- ▸ Exercises and experiments
  1. Your body language
  2. Others' body language
  3. Problems with peers: Focus on the problem, not the person
  4. Hidden peer conflicts

This chapter shows how you can apply the core skills from Chapters 1–3 to improve your ability to work through difficult situations with peers in the science workplace.

PEER PROBLEMS IN THE SCIENCE WORKPLACE

Whether you are a graduate student, a professor, or a scientist in a pharmaceutical company, chances are that a significant part of your workday is spent dealing with colleagues

95

who engage in behavior that seems inconsiderate at best and downright malicious at worst. Examples of such behavior include

- The lab member who plays loud and distracting music without regard to its effect on others.

- The postdoc who leaves a mess wherever she works. To make matters worse, she does not consider it a mess, but you do.

- An overbearing colleague who dominates group or department meetings.

These and similar types of behavior may elicit angry reactions from colleagues. Frequently, the behavior goes unchallenged. Peers of the offender may not consider it their responsibility to confront the offending party or they hope that a higher authority figure will intervene. For many scientists, their previous experiences addressing such situations have been uncomfortable or confrontational. The result is that they avoid the problem or person. Very often, the resentment escalates until a blowup occurs.

The following case study, based on an actual situation, illustrates problems that can arise when we avoid dealing with peer issues.

▸▸ *Case Study: Talk About It*

Jack was an assistant professor at Wannabe University and a major user of the department's shared two-photon microscope facility. Because operation of this device required specialized skills, Jack had assumed responsibility for training new users. Although the device was open to all department users, Jack made it difficult for others to get trained and schedule time on the device. He subjected new users to a constant barrage of criticism, and they often found it difficult to find open time that Jack had not already reserved for either his own use or "maintenance."

Several users complained that Jack constantly hovered in the background when they used the machine, ostensibly to make sure that they did not damage the equipment. Jack's presence made users feel self-conscious and constrained. He often interrupted their work on the pretext of making sure that they had not altered the alignment or calibration of the device. As a result, department members were reluctant to use the machine and avoided having their staff trained because the experience was so unpleasant, and in some cases, traumatic.

Department members resented Jack's behavior and often complained among themselves. However, no one had ever spoken to Jack about it. Susan, a faculty peer of Jack's, had complained to the department chairman. His reaction was to downplay the issues and encourage Susan and the other faculty to deal with the matter themselves. He said, "After all, you're all mature individuals; I shouldn't have to step in for a minor matter like this." Eventually, the chairman agreed to have a discussion with Jack but it focused on the details of how users scheduled time on the device. Some adjustments were made in the way time was allocated to users, but Jack's behavior remained unchanged and the problem persisted.

Susan finally went to Ellen, the head of human resources, to complain about Jack's behavior. When Ellen asked if Susan or anyone else had ever had a discussion with Jack about his behavior, the answer was "no." Ellen agreed to speak with Jack to help resolve the problem. The following day, Ellen met with Jack and asked if he had any idea of the resentment toward him regarding the way he acted relative to the instrument. Jack was genuinely baffled; no one had ever said anything to him about this. He professed to have no hidden agenda and seemed hurt that others would find his behavior objectionable without ever bringing it up to him.

KEY ELEMENTS OF PEER CONFLICT AVOIDANCE

Variants of this scenario are widespread, if not universal, in science and technical settings. One element is the scientist who behaves in a self-interested or objectionable manner, but whose behavior is not so egregious to seem overtly inappropriate. In Jack's case he never actually prevented anyone from using the device, and all of his "interventions" were couched in the context of keeping the machine in good working order so that work would not be jeopardized.

The second element is the reluctance of peers to confront the objectionable behavior. When asked why they never mentioned the effects of his behavior to Jack, they responded with, "It's not my job. The chairman should do it," "I felt that Jack could be volatile and I didn't want to become the recipient of his hostility," or "I could never figure out exactly what to say to him. It seemed like he could always deny any negative intent because on the surface everything he was doing was in the context of keeping the machine in good working order." Some used the excuse that if Jack was so oblivious that he did not see for himself the consequences of his behavior and its effect on others, then he would never be receptive to hearing about it from them.

All of these reasons for not confronting Jack's behavior are actually excuses for not knowing how to confront or even discuss the matter. The thought of dealing with the issue in a straightforward manner made everyone anxious. The fear that the discussion would end in hostility was a prime factor for many.

The third key element of the scenario is the failure of the department chair to take responsibility when the faculty could not. The chair was in the same situation as the others. He felt anxious about confronting Jack and found excuses for why he should not do it. When he finally did speak with Jack, it was about scheduling, a neutral, only loosely connected component of Jack's behavior. The abdication of responsibility by a scientist in a position of authority is a theme that we revisit in more detail in the case study found at the beginning of Chapter 9.

In this example, Jack had never been told about the effects of his behavior on others. When Ellen finally confronted him, he was baffled and then resentful that others had been talking about him behind his back. This is the fourth key element of such situations: The protagonist is often unaware of the effect of his behavior on others. Jack was behaving in a way that was motivated by anxiety about his own work, in which the two-photon device had a major role. Since Jack did not have a lot of interpersonal savvy, and no one had ever taken him to task for the consequences of his behavior, he remained oblivious to the consequences. The result was his feeling that people were conspiring against him.

With some coaching from Ellen, Susan and others learned how to inform Jack of the effects of his behavior by focusing on issues of accessibility and productivity, and without impugning Jack's motives. After a year, Jack had learned to watch both himself and others for signs that he was being overbearing or inconsiderate.

Another example of conflict aversion is when someone declares a successful resolution to a problem before it has actually been resolved. Declaring the problem solved alleviates everyone's anxiety, but guarantees that it will be reemerge. The following case study, taken from an attempted reorganization at a medical device company, illustrates this.

> ## ▸▸ *Case Study: Declaring Success*
>
> Ashok was the vice president for strategy at Mattehorn Medical Devices. The CEO asked him to oversee a complex merger and reorganization involving their electrical engineering groups at two different sites. Each group was headed by a powerful vice president of its own, each jockeying for advantage relative to the other. Although Ashok had no vested interest in the details of the reorganization, he did have responsibility for ensuring that the process was completed on schedule.
>
> As soon as the reorganization committee started to meet, Ashok saw that the two vice presidents' behind-the-scenes jockeying for position would sabotage the process. He was at a loss as to how to deal with this and felt very uncomfortable with the vice presidents' underlying animosities. With every issue that pitted the interests of them against one another, Ashok listened to the back-and-forth positioning of the two, but he made no effort to bring their underlying interests to the surface. Ashok felt that to do so would have required him to expose their seemingly irreconcilable expectations for control of the merged division. This would lead to an angry confrontation that he wanted to avoid.
>
> Ashok tried to find small elements of agreement between the two executives. He focused the committee's discussion on only these elements, eventually declaring, "We're all in agreement." Despite the fact that everyone knew that they were not in agreement, they went along with the charade because it released the tension in the room and obviated the need to deal with the difficult underlying issues.
>
> The result was that the two vice presidents lobbied Ashok and others outside of the meetings. Despite this, the reorganization committee came up with a plan that on the surface seemed to satisfy the charter of the reorganization. But in fact, the two executives were secretly dissatisfied and as soon as the formal process ended, each met with the CEO and argued for modifications to the plan that would favor them. Eventually the CEO became so disillusioned with the outcome that he cancelled the reorganization altogether, much to the detriment of the company.

Ashok saw his role principally as referee and peacemaker, largely because he was uncomfortable with the underlying competitive issues between the two vice presidents. He ignored the underlying conflict to avoid dealing with it. Although he did manage to find important elements on which they agreed, a number of more significant issues about which they disagreed were never dealt with during the planning process. At the time, everyone, including Ashok, accepted the omissions of these issues because everyone was uncomfortable dealing with them. However, in doing this, Ashok did them all a disservice.

Scientists who are prone to avoid conflict must be alert for situations in which they deny its existence to maintain the appearance of agreement. If the alternative to denying conflict is to engage in a bitter and destructive battle, denial may seem attractive at first glance. But more often than not, denial leads to an outcome such as that described above. The long-term solution is to negotiate a resolution that satisfies underlying interests. Those basic tools are presented in Chapter 3. In the following sections, we present ways to help you recognize when you are avoiding conflictual peer interactions, and how to resolve them.

TOOLS FOR DEALING WITH PEER CONFLICT

Recognize the warning signs

Recognizing the warning signs of peer conflict requires that you pay attention to your reactions. Perhaps you are hostile or resentful toward one particular colleague because of something that he does or says. If you feel that your only course of action is an angry confrontation, you do nothing because confrontations are anxiety provoking and can have negative consequences. But the longer you do nothing, the worse it gets. In the case study cited on page 96 (Talk About It), Jack's peers waited two full years before addressing the situation.

Danger signs requiring attention include feelings of resentment, fear, disappointment, frustration, or anger toward colleagues. Learn to "capture" or notice these feelings at the time that you experience them. Paying attention to your feelings may enable you to determine what it is that is bothering you. If it is something the other person does or says, or does repeatedly, think through some alternatives for discussing the matter with the other person. The sections below provide examples of such conversations, and the exercises at the end of the chapter help you to develop a better awareness of your experiences during uncomfortable interactions with colleagues.

Another warning sign of conflict is repeatedly complaining about a colleague. Notice when you do this and reflect on whether it is in your interest to continue complaining or find a way to discuss the problem with the colleague directly. The next section shows you how to have such a discussion.

Avoid accusations by addressing the problem in terms of its effect on you

This is always your best first approach for addressing behavior that is objectionable, inconsiderate, or insulting. For example,

Instead of "Susan, I think that you're being very inconsiderate by playing your music so loud all day" (accusation).

Say "It's really hard for me to concentrate with your music on so loud" (effect of behavior on you).

Instead of "Richard, you've consistently ignored all of the suggestions from my team about the project plan."

Say "My team and I feel excluded from the planning process. It seems as though none of our ideas have made it into the final plans."

Instead of "Juan, you keep cutting me off during lab meetings. I can't say anything without you interrupting me."

Say "I don't seem to be able to get my point across without being interrupted. I'd like to be able to express my ideas before I answer your questions."

The second type of statement makes the same point as the first, but in a nonaccusatory manner. The first statements may be technically true, but do they accomplish what you want? If your objective is to start an argument, then they do. Susan can deny that she means to be inconsiderate; Richard can deny that he ignored your suggestions or is being unfair. Juan may not be able to deny that he keeps cutting you off, but making accusatory statements may cause him to become defensive and deny your accusation. In each of these cases, using "you" statements conveys an accusatory stance and places the recipient in a defensive and possibly adversarial stance. Chances are that you will exacerbate the problem and create resentment.

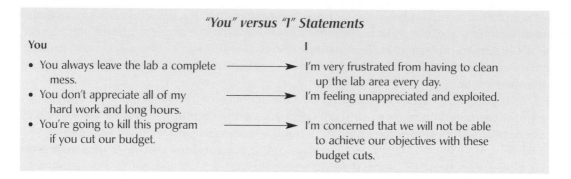

When you express your concerns in the form of an "I" statement, the other party cannot deny or refute what you say. Susan cannot deny that her music is preventing you from concentrating, Richard cannot deny that you and your team feel excluded, and Juan cannot deny that you feel that you are having a hard time making your point. Moreover, such statements present the issue as a problem that the other party can help you to solve, rather than as a behavioral deficit.

Address the specific behavior without devaluing or belittling the person

This technique is based on tools used in negotiation that teach us to be "hard on the problem and soft on the person" and is closely related to the previous technique. The difference is that, in this case, the irritating factor is not an incident but a behavior pattern. The corollary of this approach is to separate the problem, incident, or behavior from the person. I have found in my workshops that this technique baffles some scientists at first, partly because of their tendency to seek cause and effect explanations. The person seems to be the cause of the behavior, so it makes some sense, they reason, to confront the person.

While this may make logical sense, it gets you nowhere. Addressing a pattern of objectionable or unproductive behavior by implying or stating that it reflects some character defect or mental deficiency is both confrontational and threatening. The result is that the other person will not likely be receptive to what you have to say. The examples that follow illustrate this technique.

Instead of "As usual, you were really obnoxious in the team meeting today" (general accusation).

Say "Your comments about my work were unfair and even objectionable and I'd like to discuss this with you" (reference to the specific behavior and request for clarification).

Instead of "You're hogging all the time on the two-photon microscope" (attribution of selfishness).

Say "It's becoming difficult for me to find time on the microscope. Can we discuss how we schedule time on it?" (reference the specific behavior with evidence, e.g., only being able to reserve 80% of the time, and how it affects you).

Instead of "Jack, you're being possessive of this equipment and creating an atmosphere of resentment and hostility" (accusations).

Say "Jack, I'm uncomfortable when you look over my shoulder and criticize my technique. If you have concerns about the equipment, let's talk about it when I'm done here" (addresses specific behavior and its effect on you and suggests an alternative).

Defuse anger and hostility

Of course, it is difficult to use any of these techniques if the person that you are addressing is angry or hostile. In these situations, you first need to create an environment in which what you have to say can be heard. This basic approach is described in Chapter 5. You can defuse angry or aggressive behavior by using the "agree, empathize, inquire" tools. Only after you have shifted the tone of the conversation from one of hostility to one of collaboration will the other person be able to listen you. The following example illustrates how to apply some of these tools.

> **▸▸ *Case Study: Think Before You Speak***
>
> Barbara is an assistant professor in biochemistry working on the role of SMADs in signal transduction. Six months ago, she had an idea about a way to use a green fluorescent protein (GFP)-SMAD construct to monitor changes in the intracellular location of a particular SMAD during cell signaling. She discussed the idea with Mohan, an assistant professor of physiology with a lot of experience using GFP constructs. Later that year, Barbara sat in on a seminar about SMADs given by Astrid, a faculty member from another institution. Astrid discussed some preliminary work in which she used a SMAD–GFP construct in almost precisely the same way that Barbara had planned. Moreover, she indicated that this work was being done in collaboration with Mohan.
>
> Barbara felt herself getting red in the face during the seminar. She believed that Mohan had stolen her idea and she was furious. At the end of the seminar, hardly able to contain her anger, she approached Mohan and pulled him aside.
>
> "I told you that I was planning to do almost exactly that experiment six months ago. Now I find that you did the same experiment with Astrid. This is infuriating and amounts to theft of my work." During this recitation, Barbara became increasingly agitated and shouted loud enough for everyone leaving the seminar to hear, "How can you possibly justify what you did?"
>
> Mohan was stunned and embarrassed. People were looking at the two of them. In desperation, he said, "Look, you're way off base here. I never talked about your work. Why don't you calm down? You have completely misunderstood this situation and now you are making a mountain out of a molehill."
>
> To Barbara, this felt like an attempt to brush her off and she became even more furious. "You're an outright liar, Mohan, and I'm taking this to the committee on scientific misconduct," she shouted.
>
> After hearing Barbara call him a liar, Mohan became furious and said, "Go ahead. You're paranoid and everyone knows that."

In this case, we have the advantage of knowing that Barbara did in fact talk to Mohan about her idea. But we do not know whether Mohan knowingly misappropriated it or something else happened. Mohan's reaction simply fanned the flames of Barbara's anger and resulted in her filing a formal charge of misconduct against him.

Mohan did everything wrong when confronted by Barbara. He denied her anger, or that she had any reason to be angry, by telling her to calm down. Telling an angry person to calm down is probably the least effective way to get them to do that. In fact, it is likely to increase their anger. Then Mohan told Barbara that what she is furious about is not a big deal. It was clearly a big deal to Barbara, and hearing Mohan deny that she had something to be angry about did not help. What could Mohan have done differently? Let us rerun the tape with a new and different Mohan.

Barbara: "I told you that I was planning to do almost exactly that experiment six months ago. Now I find that you did the same experiment with Astrid. This is infuriating and amounts to theft of my work. I demand an explanation."

Mohan: "Barbara, I can see that you are really angry. I know how this looks to you, and I'd be angry too if I thought that someone did that to me. Frankly, it should have occurred to me how this would look to you, and I apologize for not speaking with you sooner. Can we go somewhere and talk about this? I'd really like to know what you think happened. And I'd like to explain how this situation came about. The last thing I want is for our relationship to be damaged because of this... ."

Principles used

Empathize. *Mohan acknowledges Barbara's anger and shows that he understands why she is angry.*

Agree. *Mohan has agreed that Barbara has the right to be angry based on what she thinks happened. In doing this, Mohan has not agreed that he has done anything wrong.*

Apologize. *Without agreeing that he did anything unprofessional, Mohan apologizes for his role in the misunderstanding.*

Inquire. *Mohan indicates that he wants to hear what Barbara has to say.*

Assure. *Mohan assures Barbara that their relationship is important to him.*

Mohan immediately starts by empathizing with Barbara. He tells her that he can understand her anger, rather than deny that she has reason to be angry. Note that telling Barbara that he can understand her anger is not the same as admitting that he did anything wrong. In fact, in this case, Mohan did not do anything wrong. But he needs to create a climate that enables him to explain this.

Next Mohan apologizes, not for doing anything wrong, but for failing to anticipate how Barbara would perceive the situation. Apologies work wonders with angry people, even if you are not apologizing for precisely what they are angry about. It shows Barbara that Mohan cares about her feelings and is willing to accept some responsibility.

Mohan tells Barbara that he would like to hear more about what she believes happened, further showing that he is interested in her perception. And finally, he assures her that it is important to him to maintain their relationship. All of this will allow Mohan to explain the situation from his perspective in a calm setting.

When they met later, Mohan explained that after Barbara initially came to him for advice he believed that that was the end of their interaction about the experiment. The very next month, Astrid approached him with a similar request, asking Mohan to collaborate on the project. Since Astrid was a competitor of Barbara's, Mohan felt that he could not tell Barbara about Astrid's work, and vice versa. Mohan felt in a bind. For lack of any other alternative, he went ahead with the collaboration with Astrid.

The issue here is not whether Mohan did the right thing. The issue is what the two participants can do to resolve the problem. In the second scenario, Mohan refused to react to Barbara's insults, and he created an atmosphere in which Barbara was willing to listen to what he had to say. In the first instance, his behavior resulted in a lot of people spending a lot of time adjudicating what needed no adjudication.

There is another aspect of this case worth noting. Mohan's collaboration with Astrid was not the result of maliciousness, but rather, confusion and perhaps naïveté. Recall our discussion of the fundamental attribution error. In our experience, most instances that seem as though a colleague has behaved in an insulting or malicious manner are in fact the result of thoughtlessness. Barbara was prepared to think the worst of Mohan without knowing all the facts. She owns a lot of the responsibility for the misunderstanding.

In Barbara's case, awareness of her hot buttons might have helped a great deal. Barbara's strong reaction to Astrid's presentation suggests a predisposition to feeling exploited. If Barbara had known this about herself from previous incidents, she might have been able to recognize what was happening to her during Astrid's presentation. She might have asked herself whether an angry confrontation was really what she wanted. Better self-awareness on the part of either of the participants would have prevented an ugly and unnecessary incident.

HOW TO DISAGREE ABOUT SCIENCE WITHOUT ARGUING

It is not uncommon to observe participants expressing hostile or insulting remarks to one another in a team meeting, especially when discussing their interpretations of data. Nothing frustrates scientists more than having to listen to presentation of data that they believe are flawed or just plain wrong. I have seen people become furious listening to someone recount the results of a poorly designed experiment; they squirm in their seats, make pained facial expressions, and interrupt the presentation with aggressive, snide, or hostile questions. Some scientists consider it their obligation to express their frustration overtly and in damaging ways. This was exactly what Fred thought in the case study discussed in Chapter 2.

There often seems to be an unspoken agreement that the importance of good science is so great that objections, criticisms, and destructive comments can and should be delivered swiftly and mercilessly. The corollary is that people are simply the instruments through which questionable data are delivered, and their feelings and reactions need not be taken into consideration when criticizing their results. Unfortunately, the outcome is that important comments, critiques, and suggestions get lost in the noise of hostility. Scientists who display such attitudes fail to recognize that valid scientific criticism has to be heard and accepted by the recipient to be useful. Consider the following example.

Joe says, "That binding data can't possibly be right. You didn't do it in the presence of magnesium and everyone knows that magnesium is required. You did this all wrong." Fred immediately feels attacked and gets defensive. "No, you're wrong. It doesn't need magnesium; almost any divalent cation will do. I've done it this way before and presented the results in previous meetings. You must not have been listening."

This is the beginning of an argument. Both sides now feel insulted and defensive. The ensuing discussion will likely be more about scoring debating points than determining whether the reaction really should have been done in the presence of magnesium.

The reasons for interactions such as this are multiple. Team meetings are often regarded as competitions rather than peer-to-peer discussions. This is the responsibility of the team leader to correct, and it is addressed in detail in Chapter 4. The team leader must create an atmosphere in which science can be discussed without participants feeling the need to score points at the expense of others. She must also model how to question and critique data in a way that is both scientifically rigorous and respectful of the presenter. The following are some examples of how this can be done.

Instead of "That binding data can't possibly be right. You didn't do it in the presence of magnesium and everyone knows that magnesium is required. You did this all wrong."

Say "Fred, I thought that this reaction required magnesium. That could explain why your results look different from last time."

Instead of "I can't believe that you did this experiment without a control group of untreated rats. That makes the whole experiment useless."

Say "Was there some reason that a control group with untreated rats was not included this time?" or, stronger, "I understand the results, but I'm having a hard time feeling confident about them because of the absence of a control group with untreated rats. What was your rationale for not including one?"

Instead of "I don't really understand anything that you have just said. It makes no sense to me."

Say "You're going to have to help me understand this better. I think that I missed the point of what you just said."

Instead of rolling your eyes and ignoring what you think is a lousy idea,

Say "Okay, that's one way of looking at it. It's not one that I favor, but I hear what you're saying."

To adopt this type of moderated approach first notice whether you have feelings of superiority, frustration, or aggression before you speak. If you are not aware of these feelings, there is a greater chance that you will simply react to them and say something that comes across more as hostility to the experimenter than as a critique of the experiment. Even if you are asking questions in a way that does not seem overtly hostile or aggressive to you, the recipient of your questions will probably become defensive or act threatened.

Scientists have a strong proprietary interest in their work and data, and even innocuous questions can feel threatening to them. In the best of all worlds, we would listen and respond to the content of a question and put our feelings aside. But we do not live in that world.

You can and should phrase your questions in a way that increases the chances of a substantive discussion of the issues that concern you. Reviews of experiments or data do not need to be conducted like the Spanish Inquisition. Here is another example.

Alice is presenting some disappointing results of an assay to measure the power output of a new type of light-emitting diode (LED) that her company is developing. Raul listens carefully and then interrupts to ask what seems like a straightforward question: "Alice, are you sure that you had the photometer calibrated when you measured the output?"

Alice, who has been working with LEDs for about five years, feels insulted that anyone would ask her such a question. After all, she knows perfectly well that she calibrates the photometer with every use. Alice therefore responds, "Yes, I'm sure. I've only been doing these measurements for five years, Raul." End of discussion.

What if instead Raul had said, "Alice, the last time that I made similar measurements, I had a really hard time calibrating the photometer. Could this be part of the reason for your results?"

The take-home message is that you will usually get a more productive and useful response if you phrase even innocuous questions as requests for help or clarification, rather than as challenges. Most people respond positively to requests for help.

PAY ATTENTION TO YOUR OWN BODY LANGUAGE

Reading between the lines of what people say by watching how they move, gesture, or hold their bodies can be a valuable source of information. Becoming aware of your own body language during interactions can help you to identify feelings for which you may not consciously be aware.

Watching body language while talking with someone is the best way to get feedback on how they are reacting to what you are saying. In some cases, body language may be all that you have to go on if someone is silent. However, be cautious about overinterpreting body postures or movements. Always ask questions to verify any conclusions drawn from watching someone.

You may already do this to some degree. Very often, when we speak with someone, we may notice a change in facial expression, shifting of their body, or sudden loss of eye contact. In such a situation you might stop and say, "Did I just say something with which you disagree?"

The following is a list of some common body movements or behaviors that can convey information.

Positive: Expressing interest or empathy

- Maintains good eye contact

- Faces the other person directly, rather than at an angle

- Nods head in affirmation

- Says "uh huh" as an acknowledgment

- Wears an empathetic facial expression or smile

- Leans forward when listening or speaking

Negative: Expressing anger, concern, anxiety, or lack of interest

- Gives little or no eye contact

- Suddenly or transiently loses eye contact in response to something that the other person says

- Folds legs or arms across body

- Vibrates leg

- Leans back from the other person

- Does something distracting such as pick lint from clothes, check phone for messages, furrow brow, etc.

- Does not smile or nod in recognition

Be aware that you convey your feelings and thoughts through not just what you say and how you say it, but also your body language. I spent many years sitting in meetings with a look of profound skepticism on my face that often turned into a scowl. When someone pointed out this habit to me, my reaction was, "So? That's how I feel. I'm skeptical of everything." This was indeed how I felt, and it may even be an important attribute; after all, skepticism and challenge are at the root of scientific rigor. But my facial expressions often made others defensive and inhibited communication.

It may seem obvious to you that colleagues will be much more receptive to questions or comments if you deliver them in a neutral tone of voice and with a facial expression that conveys interest and curiosity, rather than hostility and disgust. But thinking this way is a far cry from behaving this way. Practice becoming aware of what you are feeling when you are about to ask a question. If you feel angry, frustrated, disdainful, or disgusted, ask yourself if your behavior, tone of voice, or facial expression is conveying these feelings. Notice what you feel before or as you speak. It will enable you to decide how you want to come across.

None of the above implies that you must never express feelings of anger or frustration. The point is to take control of when and how you do so. Demonstrating mild anger or frustration can on occasion make your point and get the listener's attention in a way that reasoned statements cannot. But the expression of these emotions should be closely monitored and used knowingly and sparingly.

Also note that body statements that are inconsistent with the verbal message will negate your message and confuse the recipient. If you are verbally expressing support but avoiding eye contact, looking out the window, or leaning back from the conference table, you are sending a mixed message. Remember to especially monitor your body language when you need to say or do something that makes you uncomfortable, or in situations where you may not be in full agreement. Such situations are common in the workplace, where we sometimes must say or do something that is appropriate and professional but makes us uncomfortable personally.

For example, if you say, "Sara, would you be willing to collaborate with us on this project? We really need someone with a background in genetics," but you are gazing out the window, making only transient eye contact, and standing four feet from Sara, she will likely get the message that you are requesting this as a formality rather than out of genuine interest. Assuming that Sara is the best choice for this project, but you are threatened by her or believe that she dislikes you, your professional responsibility may be at odds with your feelings. Because these particular feelings are not terribly relevant to whether Sara is the right collaborator, you probably do not want to compromise your request by making one comment with words and another with your body. If you pay attention to your feelings as you approach Sara, you will anticipate that they may affect your behavior. An awareness of your facial expressions and body movements will help you to convey a consistent message. The first exercise at the end of the chapter helps you to observe and make conscious choices about your own body language.

PAY ATTENTION TO THE OTHER PERSON'S BODY LANGUAGE

While you watch your own body language, watch the other person's as well. The way in which they move, behave, and look at you may be more revealing than what they say.

Look for the behaviors listed on page 105 for clues to what is really going on, and follow those observations with questions to ensure that you are interpreting their behavior correctly. In a collegial discussion with a peer, such observations and inquiries can enable you to gauge the other person's reactions and facilitate a productive discussion. If the other person is dissembling, or saying one thing while behaving in a seemingly contradictory manner, you may have reason to question their sincerity or commitment. You need not confront them with your uncertainty, but it would likely motivate you to seek stronger assurances about any agreement. Exercise 1b on page 109 helps you to practice your skills at reading body language.

CONSEQUENCES OF FEELING INFERIOR/SUPERIOR TO PEERS

Feelings of inferiority and superiority are common in scientific settings, where performance and knowledge are constantly being tested and measured. If you are unsure of yourself to begin with, working in a competitive environment can make matters worse. Feelings of inferiority may cause you to behave in a self-denigrating or self-effacing manner and downplay your skills or accomplishments. The more you behave this way, the more you and others will believe in your limitations. The resulting consequences for your career can be unfortunate: Important contributions may be ignored, you may be passed over for promotion, and your ideas and comments may be viewed as unimportant. Your feelings end up becoming self-fulfilling prophecies. Some scientists explain their self-deprecating manner as a desire to be humble. If you think that you are being humble but you feel marginalized, ignored, and bypassed, it would be wise to reexamine your motivations and feelings.

Becoming aware of feelings of inferiority is the first step toward overcoming their consequences. The next time that you catch yourself minimizing what you know or did, remember the situation and write down as many details of the circumstance as you can. Ask yourself whether your contributions were really as limited as you had made them out to be. What would the consequence have been had you spoken in a more positive manner about yourself or what you did? Investigating your behavior in this way may enable you to uncover erroneous assumptions. For example, if you felt that speaking honestly about the value of your work would feel as though you were boasting, ask yourself why being positive about your contributions has to feel that way.

The next time you are about to do or say something that is self-deprecating, stop to think of something positive to say about yourself instead. For example, if someone says, "Alice, your role in this project was incredibly important. I don't think that it could have succeeded without you," instead of saying, "Well, others contributed as much as I did. I couldn't have done this without Mario's help—he's the one who really deserves your thanks" or "I was just doing my job; it's not a big deal," try saying, "I really put a lot into that project and I'm glad that it turned out to your satisfaction," "I was happy with the way it turned out as well," or simply, "Thanks."

The converse of being humble is to be an aggressive self-promoter. We all know scientists who constantly seem to promote themselves, their accomplishments, and their knowledge. In principle, nothing is wrong with this. Sometimes, it is what you need to do to survive and advance in a competitive world. Something is wrong, however, when this behavior is at the expense of others, the group, or team. Often, such behavior stems from feelings of inferiority or insecurity.

Just as self-deprecation can have career-limiting consequences, so too can self-promotion, when it is done in ways that minimize the contributions of others. If you are a self-promoter, it is important that you observe how others respond to your behavior. If you sense that they are reacting with skepticism or annoyance, it may be time to tone it down. Take time to reexamine how you promote yourself, paying special attention to what you may say or do that reflects negatively on others. If you need to self-promote, do it in a way that does not devalue or diminish others. Keep the focus on your accomplishments, skills, and knowledge, not on the deficiencies of others.

EXERCISES AND EXPERIMENTS

1 Your body language

Notice your body language

a. During the course of your interactions with others on a typical day, take a moment to observe how you stand, sit, hold your hands, etc. Use the list on page 105 for suggestions about what to look for. See if you can relate what you do with your body to how you feel at the moment of the interaction. Ask yourself if noticing your body language helps you to become aware of your feelings at that moment. Awareness of your own body language can alert you to how you might react in a future situation and gives you the opportunity to consciously choose your behavior.

b. Write down several instances of your body language messages that are consistent and inconsistent with what you are saying. Review each situation and ask whether the interactions in which your body language was consistent with your message were more productive than the others.

c. Make a list of three people with whom you have significant interactions at work in a typical week. For each of these people, the next time you interact, record the following responses:

	Agree	Disagree
1. I am uncomfortable making eye contact	☐	☐
2. I feel tense	☐	☐
3. I fidget	☐	☐
4. I cross my arms or legs	☐	☐
5. I routinely make eye contact	☐	☐
6. I stand close to the person	☐	☐
7. I behave in a relaxed manner	☐	☐
8. I am comfortable touching or making other appropriate physical contact	☐	☐

The people for whom you agreed with statements 1–4 are probably those with whom you have some discomfort or uneasiness. Ask yourself whether this uneasiness is affecting your work or professional relationship in a negative manner.

d. Pick one person from the "agree" list of statements 1–4 as an experimental subject. During your next several interactions with this person, make it a point to use one or more of the positive body language behaviors listed on page 105. After the interaction, ask yourself the following questions:

• Did my changed body language affect the dynamic of the interaction?

• Did it change the way in which I interacted with the other person?

- Did it change how the other person responded or acted toward me?

- Did these changes lead to a more productive outcome?

2 Others' body language

List three people with whom you interact during a typical day. For each of them, watch and record any of the body language messages listed on page 105 during your interactions with them for one day. Does taking note of these messages give you information about your interactions? Write down several instances of body language messages that you observe that are consistent and inconsistent with what the person is saying. Often, a person's body language betrays their discomfort with what they are saying. In one or more cases where body language seemed inconsistent with what was being said, ask questions that encourage the other person to elaborate, clarify, or restate what they said. Does this technique improve the quality of your communication? Does taking note of body language help you to have a more productive interaction?

3 Problems with peers: Focus on the problem, not the person

In Chapter 3, we introduced the concept of focusing criticism or frustrations on problems, not people. This concept is, of course, applicable in almost any situation, and we use it liberally throughout the book. The following exercises can help you apply this principle to interactions with colleagues. Write down your answers to each question. Sample answers for situations 1 and 2 follow.

Situation 1. You and your fellow faculty members have agreed to set aside a room in your research building to be shared by all. However, when you return from a three-week vacation, you discover that one of your colleagues has installed a mass spectrometer in the room as well as two of his technicians, effectively taking up the entire space.

- What would happen if you do nothing?

- What could you say that could be construed as attacking your colleague?

- What could you say to attack the problem?

Situation 2. After having a confidential discussion with a colleague, you discover that someone from another institution knows the details of your discussion.

- What would happen if you do nothing?

- What could you say that could be construed as attacking your colleague?

- What could you say to attack the problem?

Situation 3. Describe a current conflict that you are having with someone in your workplace. Answer the following questions about how you might deal with this problem.

- **Avoidance.** What would you do if you avoided the problem?

- **Attacking the person.** How would you address this problem by attacking the person?

- **Attacking the problem.** How could you focus on the problem?

    List as many of your underlying interests as possible for resolving this problem. List the consequences of each of the three approaches. Which approach enables you to satisfy the greatest number of your underlying interests?

Situation 1 sample answers

- **Avoidance.** You do not say or do anything. The result is that everyone feels resentful and that group agreements can be ignored at will.

- **Attacking the person.** "What do you think that you're doing in the equipment room? You've hogged all the space and left none for the rest of us. This is inconsiderate and outrageous."

- **Attacking the problem.** "I thought we agreed that this space would be shared. Was that your understanding? The situation as it stands now makes it impossible for the rest of us to use the room. Perhaps we need to ask the group to review our understanding of the agreement to see how all of our needs can be accommodated."

Situation 2 sample answers

- **Avoidance.** You do not say anything and you never discuss your work with him again. You miss the benefit of his expert advice but it is not worth the worry. This results in a delay of your work.

- **Attacking the person.** "I can't believe that you talked about my project with the people from MIT. Have you ever heard of confidentiality? I thought that I could trust you, but that's clearly not the case."

- **Attacking the problem.** "We seem to have had a misunderstanding about our communications. I was under the impression that our conversations were confidential, but it seems as though you were under a different impression. Am I right about that? If so, I need your assurance that this won't happen again. If you cannot agree to that, I'll have to discuss my work with someone else. That will really delay the progress of this project, but I feel as though I have no other choice."

4 Hidden peer conflicts

Many of the difficulties that we have with peers stem from feelings that have nothing to do with the person herself, may be partly our own fault, or are based on lack of familiarity or contact with the person. Unless we are aware of these feelings and how they influence what we say or do, we are at their mercy. If you are conflict averse, you probably avoid thinking about the problems or uncomfortable feelings associated with these people. Remember that you do not need to deal with every problem that you identify. Focus on those that you think will have the greatest impact on your work and your organization. The following exercise will help you to become more aware of hidden or ignored conflicts that may be interfering with your work.

    Make a list of as many colleagues as you can with whom you have a problem, conflict, or just uncomfortable interactions. For each person, answer the following questions:

a. What problem do I have with this person?

b. Is the problem interfering with my work?

c. Is the problem interfering with my responsibilities to my organization?

d. Even if it is not interfering, could my work be improved if the relationship were improved?

e. What could this person do to alleviate the problem?

f. What could I do to alleviate the problem?

g. How could I alleviate the problem without attacking the person?

CHAPTER 7

The Slings and Arrows of Academe: Survive to Get What You Need

This chapter shows how trainees can cope with the sometimes confusing and stressful interactions they can have with their mentors. We also show how mentors can provide training experiences to help trainees, as well as themselves, be more productive. Chapter 8 applies these same concepts to managing in the private sector, which has its own unique set of problems.

We begin this section of the book with an exploration of science management in the academic setting since many poor management practices in the science workplace originate there. The standards, mores, behaviors and expectations learned in academe remain with scientists throughout their careers. Scientists in training may become imprinted with their mentors' management styles, perpetuating poor management from one generation to the next and from academia to the private sector. Although many academic mentors are skillful leaders, just as many, if not more, are not. The following examines why academic institutions pay no heed to managerial and interpersonal skills in their faculty and what can be done about it.

- ▶ **Academic research: A fount of creativity or a den of dysfunction?**
  Academia rewards individual achievement
  The dark side of the academic reward system
  Who is in charge here?
- ▶ **An ill-fated academic research project**
- ▶ **What went wrong?**
- ▶ **Improving your skills as a mentor**
  Trainees learn from your behavior
  Anticipate the consequences of your behavior
  Focus on process, not just content
- ▶ **Surviving as a trainee in academia**
  Observe how your mentor manages and relates to trainees and employees
  Be aware that your mentor may be oblivious to the impact of his behavior
  Take account of the pressures and deadlines to which your mentor is subject
  Your needs are not always synonymous with those of your mentor
  Let your mentor know the impact of his behavior without being accusatory
  Remember that you are not alone
- ▶ **References**
- ▶ **Exercises and experiments**
  1. Mentors: Experiments with new behaviors
  2. Trainees: Experiments with new behaviors

ACADEMIC RESEARCH: A FOUNT OF CREATIVITY OR A DEN OF DYSFUNCTION?

Your time in academia can be the best of times and the worst of times, and usually, both. On the one hand, you have the freedom to engage in exciting research with intelligent, motivated people. On the other hand, you may end up feeling like an indentured servant with back-stabbing mates, working under a mercurial master for a pittance. Scientists in academia are demeaned, humiliated, exploited, and bullied more than at any other time in their careers. How they deal with these experiences determines whether they survive to complete their training.

The academic laboratory is arguably today's most successful paradigm for scientific discovery, technology development, and engineering advances. Such laboratories are run by scientists who are sometimes referred to as "principal investigators" (PIs), whose research is supported almost exclusively by competitive research grants from the government or private foundations. For the majority of PIs, the quest for funds to support their research is never-ending, time consuming, and a source of considerable anxiety.

Grants mean money, and money translates into the ability to hire scientists and technicians, buy equipment and research supplies, and attend scientific meetings. Difficulty in obtaining a grant or in renewing an existing grant may jeopardize your job. Thus, academic scientists are highly motivated to generate funds to support their research. But this motivation is based not only on the negative consequence of losing one's job, it is also based on the positive reinforcements of obtaining research funds.

Academia rewards individual achievement

The rewards from academic research include professional prestige, recognition, and funding. These materialize largely from outside the scientist's own organization in the form of grants, publications, and professional recognition. Scientists are also rewarded by their own institutions through promotions and other perquisites (space, departmental funds, etc.) that are allocated largely on the basis of the scientist's external recognition. Thus, there is a strong and direct connection between the output of the PI's laboratory and the rewards that accrue to the PI. Moreover, because PIs correctly perceive the connection between what they do and the output of their lab, they behave like entrepreneurs, working for themselves, albeit still a part of a larger organization.

PIs are not the only ones to benefit from their hard work and entrepreneurial drive. It is also in the interest of the institution, dean, and department chair that their PIs are recognized and rewarded. Productive scientists mean more funding for the institution, prestige that facilitates fund-raising, and political leverage for deans and department chairs. For all of these reasons everyone roots for the success of the faculty scientist.

The motivation of members of the PI's laboratory is also easy to define. The trainees clearly expect that their performance will be noticed, recognized, and documented (eventually in the form of professional publications), and that this will have a direct bearing on their future opportunities.

The dark side of the academic reward system

Reward for individual achievement is at once the strength and the bane of academic institutions. It is a strength because achievement is tightly coupled to a powerful and valued system of rewards that confer professional satisfaction, recognition, prestige, power, and advancement. Yet this efficient reward system presents significant challenges to the academic system. Because every research laboratory succeeds or fails based on its accomplishments, the interests of academics are largely self-interests. Motivations to contribute at a departmental or institutional level are of a secondary nature. Chairing committees, interviewing prospective graduate students, and even teaching are viewed more as responsibilities to be fulfilled than as opportunities to help advance the individual, the department, or the institution. Moreover, because the coins of the academic realm are funding and publications, department chairs and deans are at a loss when trying to reward communal service. Thus, although the academic research enterprise is characterized by high creativity, motivation, and entrepreneurship, it is challenged by a paucity of group cohesion and an absence of shared objectives and common focus.

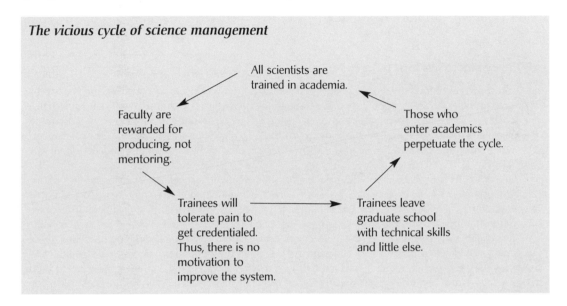

The vicious cycle of science management

All scientists are trained in academia.

Faculty are rewarded for producing, not mentoring.

Those who enter academics perpetuate the cycle.

Trainees will tolerate pain to get credentialed. Thus, there is no motivation to improve the system.

Trainees leave graduate school with technical skills and little else.

Who is in charge here?

Because academic institutions benefit from their achievements, faculty in general and scientists in particular are given wide berth for conducting their affairs and running their laboratories. Only in the most egregious cases of sexual harassment, research misconduct (e.g., falsification of data), or financial wrongdoing will a university or departmental official step in or meddle with the way in which a PI runs his or her laboratory. An almost universal lack of oversight of, or even interest in, the management styles and interpersonal behaviors (within bounds) of PIs exists in academia. The result, and I speak here from personal observation and experience, is that management of many academic labs is characterized by bullying, insensitivity, threats, incompetence, and neglect. I have seen these characteristics in labs run by scientists who were successful by any of the conventional criteria, and in labs run by scientists whose careers were foundering.

This void in oversight can also be attributed to the fact that those in a position to encourage improved management and interpersonal practices—department chairs and deans—are often themselves scientists who lack the very skills that need to be promoted.

If you are a trainee associated with a mentor who has poor managerial and mentoring skills, the chance that someone in your institution will either notice or do something about it is minimal. If you are a mentor, and you find yourself challenged by your responsibilities to trainees, the chance of your getting sage guidance from the leaders of your organization is also minimal. In both cases, your best option is to learn to observe, assess, and improve your own behavior.

An ill-fated academic research project

The following case study illustrates issues that can arise between mentors and trainees in an academic setting. The case also illustrates how both scientific and social factors play major roles in the outcome of scientific projects. Following the case, we review what happened and what could have been done differently to improve the outcome. The case is a composite taken from my own first-hand experience and from the experiences of others from whom I have heard first-person accounts.

▸▸ *Case Study: What's Wrong Here?*

Liz is an associate professor of Biology at Wannabe University working to discover the mechanism by which certain types of white blood cells move through the body to seek out and engulf pathogens. Fred, one of her research fellows (a postdoc), has generated data demonstrating the importance of a previously unidentified protein in the movement of white blood cells. Fred presents this data at the weekly lab meeting during which the Liz and her six postdocs, two technicians, and two graduate students review the week's work.

One of the technicians and Sue, the postdoc for whom she works, meets Fred's data and his presentation with disinterest because they have just started an experiment of their own. They are exchanging written notes about when the next sample needs to be taken off ice and placed in the 37° incubator. Fred notes this with mild annoyance because Sue knows a lot about his project and he had hoped for her feedback.

Two of the other postdocs with the most experience, Alan and Richard, ask probing questions about Fred's data in a way that makes him feel defensive. He feels as though he is being given a hard time and that if someone else were presenting the data, they would be readily accepted. Moreover, Alan and Richard are close friends and frequently challenge others over seemingly minor scientific issues (this seems like one of those times to Fred). Because he is distracted by these feelings, Fred fails to take careful note of all of the questions or objections raised about his work. After the meeting, he decides to avoid too much discussion about the most controversial part of the project again because he is fed up with the nit-picking of his data. He decides to quietly proceed with that portion of the work on his own.

Liz is intrigued by the data and immediately decides that she would like more effort directed to the new protein to discover its function more rapidly than Fred could on his own. She also begins to consider how Fred's work could be used in a grant application that she is completing. Because she is so preoccupied with her thoughts, she stops asking questions. Fred interprets this to mean that she agrees with Alan and Richard's criticisms, and has also lost interest in what he is saying.

After the lab meeting and without any preliminary discussion, Liz calls a meeting with Fred and a more junior postdoc, Zheng. Liz says that she would like Zheng to get involved in the project to accelerate the work. Fred is taken by surprise. After the criticism that he felt during the group meeting, he feels that this is an attempt to double-check his work or perhaps pry it away from him. He is so angry that he mentally resolves to start looking for another position as soon as possible, something he has been considering for a while. The next day, Zheng tells Liz that he has his own project and does not wish to dilute his effort by working on someone else's project. Liz then tells Harvey, her most junior postdoc, that he must work with Fred. It takes several months to train him because Fred is unenthusiastic and is spending time looking for a new position.

Eventually, Fred has found another job in another laboratory. Liz assigns Harvey as lead investigator of the project. During a lunchtime discussion with a colleague, Liz laments the amount of time lost on this project and her concern about her grant application. The colleague notes that another professor, Lee, has a postdoc who has done similar work and may agree to collaborate on the project. Liz says that she does not think this would work since she and Lee had a contentious disagreement several months ago.

After lunch, Liz wonders why so many things have gone wrong in this project, but she cannot seem to put her finger on any one crucial incident. Delays and impediments seem to have cropped up time and again. Maybe, she thinks, it is just bad luck.

WHAT WENT WRONG?

If you have worked in a research lab, you probably will not find anything in this case study to be out of the ordinary or surprising. Liz is not a malicious or inconsiderate person. She is a dedicated scientist trying to be successful and train her staff and students. Yet at almost every juncture, progress was either delayed or hindered by an interpersonal, not scientific, issue. Here are a few of the lost opportunities and their consequences.

Incident	Alternative behavior	Possible effects of the alternative behavior
Liz becomes distracted during Fred's presentation.	Liz may know that she has a tendency to become distracted and distant because this is probably not the first time that this has happened. She could have told Fred what was on her mind so that he could adjust to her new level of interest.	Fred may not have felt dismissed by Liz's apparent distance and might have been able to better focus on feedback during the meeting.
Sue, the senior postdoc, shows little interest in Fred's presentation.	Liz often multitasks at lab meetings. She brings her computer and reads her e-mail during presentations. Liz has set a precedent that it is acceptable to pay only peripheral attention when someone speaks. She could become aware that this behavior has a negative impact on her lab members. She could change her behavior to show them that it is important to pay attention during presentations.	Lab meetings could become more interactive and productive. Presenters such as Fred would not leave feeling that the meeting was a waste of time.
Alan and Richard gang up on Fred.	Liz could have moderated this discussion, instead of becoming distracted by her own thoughts. If Alan and Richard have a record of double-teaming lab members, it is Liz's responsibility to call attention to their behavior and teach them to raise their concerns in a more productive manner.	Alan and Richard may have asked important questions that would have been more thoroughly discussed in a nonconfrontational atmosphere.
Liz springs Zheng on Fred with no warning.	Springing Zheng on Fred without warning showed a lack of awareness of how he would feel and react. Liz needed to prepare Fred by explaining that she wished to help him, not take the project away from him.	Zheng's participation could have helped Fred move the project forward. Instead, a more junior person got assigned, with detrimental consequences.
Liz fails to take advantage of Lee's postdoc because of an unresolved conflict with Lee.	Had Liz been proactive in recognizing and resolving conflicts with her peers, she might have had a better relationship with Lee and gotten the help that she needed.	Liz would have been able to recruit expert help for the project, further accelerating the work.

This case is a good illustration of the complex interplay between technical execution and social dynamics in the lab. By being more attentive to her own behavior and its impact on others, Liz could have changed the outcome of this story. She might have taken steps to defuse animosities and misunderstandings in the lab before they ended up hampering the research. She also could have mended her relationship with Lee, enabling the participation of others in the project. Progress was hampered not by scientific, but by interpersonal matters.

The case contains examples of interpersonal interactions that most working scientists will find unremarkable and even routine. No overt battles took place, no threatening confrontations occurred, and no one stormed out of a meeting in a huff. No one behaved in a reprehensible or censorious manner. Yet the interactions in aggregate resulted in major delays in an important project, lost professional opportunities for the participants, and wasted research dollars.

All of the scientific participants were negatively affected by the outcome. Fred left the lab before key studies could be completed and failed to gain senior authorship on a major scientific paper. Harvey ended up owning a project prematurely, prolonging the time it took to generate important data. Liz, under pressure to generate this data for a grant renewal, failed to do so. Her renewal was turned down, and it took another year to produce the results needed to resubmit the application. The National Institutes of Health ultimately paid for several years of work on a project that yielded fewer results than might have been possible had it been managed better.

The antecedents of the outcome are many. Liz never received any training in managing a group of scientists. Her primary objective was to retain her funding, and she viewed her students and postdocs as instruments for meeting this objective. Sue's department chair did not know the way in which Sue ran her lab and offered no help or advice, despite the fact that he had invested several hundred thousand dollars of departmental funds in Sue's research. Sue's colleagues may not be any better equipped than her at discerning the reasons for the misunderstandings, interpersonal clashes, and poor management skills that hampered her work.

IMPROVING YOUR SKILLS AS A MENTOR

There are many reasons why observing, evaluating, and improving your skills as a mentor* should be a high priority if you are in the sciences. First and foremost is the fact that you have a responsibility to your trainees to provide them with the kind of professional guidance that only a mature and experienced professional can provide. The many facets of good mentoring are discussed at length in numerous excellent guides and publications. The basic functions of a mentor in a scientific setting are reviewed by Barker in *At the Helm* in the chapter on mentoring (Barker 2002, pp. 181–230), which also contains a comprehensive list of readings and references relative to mentoring in the life sciences. Skills discussed by Barker include helping trainees to establish a reliable methodology for their research, acquire scientific oversight and depth, and model appropriate behaviors for interactions with peers and collaborators.

---

*For purposes here, mentor also refers to someone who oversees a trainee's lab work or research.

Some of these functions involve transmitting information, whereas others are more subtle and involve modeling behaviors and attitudes. It is this latter group on which we wish to focus here. In the absence of well-developed self- and interpersonal awareness, you may find yourself inadvertently transmitting ineffective attitudes and behaviors to your trainees.

Another reason for giving a high priority to improving your skills as a mentor, especially in academia, is that your productivity as a principal investigator is directly connected to the productivity of your trainees. If you behave in ways that alienate or confuse trainees, you may decrease their productivity in the lab, thereby sabotaging yourself and your own interests in the process.

Trainees learn from your behavior

Trainees pay close attention to what mentors say and how they behave. Trainees may interpret overt behaviors such as belittling and maligning colleagues as the norm if a mentor exhibits them routinely. Similarly, passive or covert behaviors such as ignoring conflict and paying no attention to the effects of your behavior may also be seen as acceptable if exhibited by a trusted mentor. None of us is perfect, and we all exhibit behaviors that we regret in retrospect, or that we would rather not have a trainee emulate. You can use your own lapses in interactions with colleagues or trainees as lessons for your trainees, just as you would use scientific triumphs and failures as learning tools. Doing so implies an acceptance of the fact that your role as mentor extends well beyond the transmission of technical information and skills. By showing trainees that you have the capacity to examine and learn from your own behavior, you transmit one of the most valuable lessons that a mentor can provide.

As illustrated in the table on page 117, the outcome of the case study could have been improved if either Liz or Fred had been more self-aware. As a mentor, Liz needed to pay more attention to the effects of her behavior on those in her group. For example, she needed to understand that trainees place a great deal of importance on what may seem like inconsequential comments, behaviors, or facial expressions.

Anticipate the consequences of your behavior

If you are a mentor with a number of people working for you and you have difficulty spending time with each of them, you may be especially vulnerable to being misinterpreted. The absence of "face time" with you will incline trainees who need feedback or attention to read more into the brief interactions that they do have with you than they might otherwise. Maintaining an awareness of your own state of mind is the first step in knowing how your thoughts and feelings will influence your behavior with trainees.

If you feel pressured, you may know from past experience that you tend to act in a distracted or dismissive manner. If you are elated about an exciting result of one project, you may ignore other projects, leading trainees to wonder whether they are responsible for your lack of interest. If your personal life is a shambles, you may be venting your frustrations on your lab without realizing it. You may be so preoccupied that your lack of interest causes work to slow down. If you are aware of your state of mind, you can anticipate behaviors that impact others and take steps to act differently. The exercises at the end of the chapter may help you to improve your behavioral self-awareness.

Focus on process, not just content

Scientists who are deeply involved in problem-solving may make comments that they realize in retrospect were ill considered, inappropriate, or hurtful. As a mentor, sound a mental alert when you become deeply involved in a science discussion with trainees. I know from personal experience that it is possible to monitor your affect and behavior without damping your enthusiasm for the science. It is important to be aware of how trainees react to you and what you say. To avert misunderstandings, watch for signs that someone has misinterpreted or had a negative reaction to what you have done or said. Students or trainees may be confused or baffled by your comments, misunderstand instructions, or disagree with your viewpoints. Yet, for any number of reasons, they may not say or do anything in response. For example, they may find you intimidating, be so unsure of themselves that they are terrified of asking for clarification, or come from a culture that frowns on questioning a mentor. In such cases, the mentor must be alert for behaviors that manifest feelings of discomfort, confusion, or anxiety.

A subtle furrowing of the brow or a brief look of confusion can be revealing. Perhaps while you are talking, you notice that your student is furiously drawing pictures of tiny bugs crawling across a page in his lab notebook. In these cases, you might ask, "Jatinder, is there something I just said that's confusing to you?", "Marie, you look like you have a question. Do you?", or "Doug, do you want to talk more about this after the meeting?"

The benefit of such attentiveness is obvious. If students or trainees are unclear about or disagree with something you said, or if they misunderstood you, it is likely that they will end up doing something other than what you expected.

SURVIVING AS A TRAINEE IN ACADEMIA

Those not familiar with the politics of academic research laboratories might be temped to wonder why trainees would choose to work in laboratories that are poorly managed, or for PIs who behave in insensitive and churlish ways. The answer is simple. Most trainees chose their mentors based on their scientific interests and accomplishments rather than on their managerial or interpersonal skills. Despite exhortations by friends and advisors to seek mentors who will be mindful of the trainees' needs, often what really matters is the mentor's research track record and prominence in their field. As a result, mentors with brilliant scientific minds but poor or even disastrous managerial or interpersonal skills can flourish and produce students in their own image. Once a decision is made to pursue a doctoral level research program, academic trainees are, for all practical purposes, indentured to their mentors.

Trainees have much at stake in this relationship. Most important is that it will lead to a granting of the credentials needed to practice their chosen profession. As a result, trainees have an intrinsic and powerful motivation to persevere and perform even if they are ignored, mistreated, or manipulated. Trainees will put up with a great deal because their objective is clear and their success depends on the mentor.

If you are a trainee, once you have spent four or five years learning both science and, by observation, the management of science from your mentor, you will take away more (or less) than what you bargained for. If you continue your career in academia, you may perpetuate the same managerial deficits as your mentor. Although this may negatively impact your productivity, it is unlikely that anyone will ever call your attention to it. If, however, you go into some other line of work, you will likely run into problems.

Scientists who enter the private sector typically learn too late that management skills have a major impact on long-term promotion opportunities. This is in stark contrast to their experience in the academic setting, and this shift in expectations is rarely, if ever, made explicit. It is a distinction that young scientists miss at their peril.

As a trainee in science, there may not be much you can do in the short term to create wholesale change in the academic world. But there is much you can do for yourself to minimize misinterpretations of your mentor's behavior. You can also learn to distinguish management styles that are worth emulating from those that are not.

Observe how your mentor manages and relates to trainees and employees

It is easy for a student or trainee to assume that successful mentors know how to manage scientists. After all, there must be some reason that your mentor arrived at where she is today. But perhaps she has her position despite, rather than because of, her managerial and interpersonal skills. Maybe if her managerial skills were better developed, she would have been even more productive and more successful.

Our recommendation is that you learn to observe management styles early in your career. The way to learn is to note the management techniques that seem to be effective. We define effective management techniques as those that help meet the objectives of the enterprise (lab, group, company) and that are implemented in a manner that is respectful of others and mindful of the consequences for the members of the organization.

Note that this definition does not mean that everyone will always be happy with every decision. Do not fall into the trap of believing that mentors or managers who make decisions that lead to disappointment or disagreement are poor managers. Effective managers often need to make difficult and unpopular decisions. Conversely, do not be fooled into believing that a lab is well managed just because everyone seems happy, and everyone including the PI goes out for beer together and on weekend canoe trips. I once knew a PI who ran this type of lab, but his graduate students typically took two years longer than most to complete their degrees.

One way to assess the quality of a mentor's decisions or actions is to ask yourself the following four questions:

1. What is the objective underlying the decision or behavior?

2. Is the objective known or made known to those affected?

3. Is the objective consistent with the mission of the lab/group/individuals?

4. Was the resulting decision or behavior implemented with respect and consideration for those affected by it?

If you know the answer to question 1, you are at least in a position to understand the rationale for a decision. Decisions or behaviors for which the objective is either unclear or unknown are bound to cause confusion and consternation. If the answer to questions 2 and 3 is "yes," then both the objective and its relationship to the (presumably shared) goals of the group are known. Decisions or behaviors that lack a clear relationship to the goals of the enterprise or its members will be resisted, resented, or ignored.

Question 4 addresses the area in which most managerial actions fall short. Managers often believe that their work is complete once their staff or trainees understand the rationale behind a decision. The result is that decisions get implemented without careful consideration of how they will impact the individuals. Mentors who make decisions that have a clear rationale, consistent with the group's mission, but who implement them in ways that result in bruised egos, hurt feelings, and alienation should be observed but not emulated.

Learn to take note of your mentor's actions and the outcomes of those actions. By evaluating what you observe as described above, you can identify those managerial behaviors that are worth emulating.

Be aware that your mentor may be oblivious to the impact of his behavior

If your mentor has limited managerial and interpersonal skills, it is likely that at some point you will feel hurt, insulted, or misused by him. Trainees often feel insecure, especially during their early years of training, and readily interpret a dismissive comment or lack of interest as being due to something they did or said. This was what Fred did when he misinterpreted Liz's distraction for skepticism or lack of interest (see What's Wrong Here? on p. 116). A good rule to follow in such situations is what I call the "95% rule": 95% of anything anyone does or says in your presence has nothing to do with you. Although mentors often have important messages and advice to give their trainees, remember that not everything your mentor says or does is a direct reaction or response to you.

It is difficult to devise a foolproof guideline to help you to decide when a comment or action is meant to convey a message, is simply a manifestation of thoughtlessness or distraction, or has nothing at all to do with you. When in doubt, you will usually be better off inquiring if there is a problem rather than worrying about whether there is one. For example, saying, "I noticed that you didn't have much to say about my project report this morning. Was there some reason for that?" is much better than spending the rest of the week worrying that your mentor thinks your work is a disaster.

Take account of the pressures and deadlines to which your mentor is subject

In the case study, Fred was either unaware of or oblivious to Liz's professional life. Keeping in mind that Liz had a difficult impending grant renewal and was eagerly seeking ways to enhance her chances of success might have alerted Fred to the possibility that any new or exciting data would be welcomed.

The pressures to which a mentor is subject should not excuse insensitive and manipulative behavior. But trainees who are aware of their mentor's pressures will be less likely to interpret distraction as disinterest and aggressive questioning about a key experiment as a personal attack. This point is closely related to the subject of the preceding section: A lack of awareness of your mentor's needs, motivations, and pressures makes it more likely that you will seek personal explanations for her behavior.

Pay attention to your mentor's pressures, as well as the needs and motivations that affect her actions. Be especially aware of grant deadlines, manuscript submissions, acceptances and rejections, upcoming tenure decisions, and other academic responsibilities such as teaching and committee assignments. None of these factors excuses bad management, but being aware of them can help you to place your mentor's behavior in context and reduce the chances that you will misinterpret the occasional thoughtless comment or lapse in attentiveness.

Becoming more aware of the world in which your mentor functions allows you to improve your relationship with her (see Chapter 5). If improving your relationship results in better guidance, more thoughtful input, or more face time, the time that you spend doing so will be well worth it.

Your needs are not always synonymous with those of your mentor

In the best of all worlds, your mentor's successful research and publishing of scientific papers benefits you as much as they do her. But at times, your mentor may have a different agenda than you. Keeping highly skilled graduate students and postdocs in the lab for long periods of time to support new work may be of great benefit to a mentor but of little or no benefit to the trainee.

It is not uncommon for trainees to feel that their own interest in moving on, either to new projects or a new lab or institution, is being inhibited by an advisor who is attending to her own interests over those of her trainee. Making your own interests and needs known (by using the types of statements illustrated below) is always your first line of approach in such situations.

Let your mentor know the impact of his behavior without being accusatory

Empathy and understanding will take you only so far with an inconsiderate mentor. Remaining silent about feeling abused, mistreated, or ignored will only increase your sense of powerlessness and frustration and provide little help in getting the advice and guidance you need. Mentors who have poor interpersonal or managerial skills or create unpleasant work environments are almost never malicious—they are simply clueless. The key to getting what you need, whether it be more face time, scientific guidance, or career assistance, is to ask for help without being accusatory. When asking for help, depersonalize sensitive topics by remaining focused on improving your work and performance. You can also depersonalize sensitive topics by using "I" statements instead of "you" statements. The table on page 124 lists examples of approaches for making your needs or feelings known.

Remember that you are not alone

Sometimes your best efforts at open mindedness and at understanding your mentor's problems are not enough. Unfortunately, some advisers and mentors are, on occasion or

Situation	Passive	Confrontational	Neutral
1. You are being ignored during meetings.	Say nothing and/or withdraw.	"No one ever pays attention in lab meetings when I present my data."	"I would be more productive if I had everyone's attention and input when I presented my data."
2. You feel that your project is being given to someone else.	Say nothing; feel resentful.	"I don't understand why you assigned Zheng to help on this project. I started it and I want to finish it."	"I have a legitimate interest in maintaining my role as lead investigator on this project. I'd have a better understanding of what you are trying to achieve if we could discuss why Zheng needs to be involved."
3. You are unable to schedule time with your mentor.	Say nothing; complain to others.	"The abstract is late because I have been unable to talk with you. You're always away or busy with something else."	"I'm sorry that this is late. In the future, what would be the best way to schedule time with you when you're busy?"

even consistently, inconsiderate or unfair. It is easy to feel that you are at the mercy of your mentor because so much is at stake in your relationship with him. Also, unless the mentor's behavior is so egregious that it falls outside of the bounds of laws or institutional policies, it is not always obvious to a trainee what is acceptable and unacceptable behavior on the part of a mentor.

As a trainee, you need to be aware of the possibility and consequences of feeling isolated and at the mercy of your mentor. In the worst case, if you feel inadequate or unappreciated, you may find yourself feeling depressed and hopeless with nowhere to turn.

Thankfully, most graduate training programs are organized such that trainees are assigned to a committee of faculty members who monitor their progress periodically. Some training programs are now being improved by creating more supportive environments for trainees. For example, Harvard University's chemistry department has established multiple approaches for providing graduate students with more opportunities for faculty interaction as well as psychological support. It is unfortunate that this was done after two graduate student suicides rather than before (for a discussion of these events, see Hallowell 1999).

Take advantage of your committee mentors. In all likelihood, you will encounter at least one faculty member with whom you can establish a rapport. In most cases, simply having someone with whom to discuss problems is all that you will need. In other cases, an ally will be important if you are being treated in an unprofessional or malicious manner.

Do not overuse your faculty ally. It is easy for an ally to suspect that you are chronically dissatisfied if you use your relationship to grouse about minor matters that have more to do with you than your mentor (see above for hints on identifying such issues). Discussing these matters with a peer, especially if she is familiar with your mentor, is a good way to validate concerns that you may have about your mentor.

Yet scientists may find it difficult to reveal their insecurities to peers, and even more difficult if these insecurities derive from being belittled or harassed by their mentor. It is easy for a trainee to feel ashamed, embarrassed, or inadequate as a result of such treatment, especially if they believe that they truly are inadequate. I have heard others characterize the tendency of scientists to keep matters such as this to themselves as being due to a "macho" culture in science, and there is likely some truth to that. This attitude is fostered by the training that young scientists receive, which, as we have previously noted, is geared toward their functioning as independent investigators in a competitive environment.

It is difficult to break the mold and seek support by opening up to a peer. But if I have learned one thing from my workshops for scientists, it is that everyone has the same problems. When participants in my workshops cite examples and problems with which they are dealing, I see others nod in recognition as one problem after another is discussed. Scientists who have rarely, if ever, discussed their problems, perhaps because they are ashamed of them, or believe that the problems are unique to themselves, suddenly see that everyone has the same difficulties. Keep this in mind. Chances are that if you share your uneasiness, fears, and concerns with a peer who seems receptive, you will find yourself hearing the same from them.

If you can avail yourself of such peer support, do not fall into the rut of using your peer group only for grousing. Use the guidelines in the sections above to critically examine and evaluate your mentor's behavior. Consider motivations, external pressures, and constraints. Use your discussions to put yourself in your mentor's place and ask what you would realistically do in the same circumstances. In many cases, you will be surprised to discover that decisions that seemed unfair or capricious were actually the result of a complex set of circumstances for which there was no easy solution. Perhaps a decision that could have been implemented with finesse and consideration was instead carried out in a heavy-handed and inconsiderate manner. Your training experience and your own success as a mentor and manager will be improved immeasurably if you and your peers use these and other types of decisions and behaviors as learning tools.

In addition to taking advantage of peer support, determine whether your institution has resources or personnel to support trainees. For example, the University of Michigan Rackham Graduate School publishes a comprehensive and insightful guide to mentoring and being mentored (http://www.rackham.umich.edu/StudentInfo/Publications/StudentMentoring/mentoring.pdf). In addition, several online resources specifically focused to helping trainees think about and deal with mentoring issues are available (e.g., http://gradschool.about.com/cs/aboutadvisors/a/mentor.htm).

REFERENCES

Barker K. 2002. *At the Helm: A Laboratory Navigator.* Cold Spring Harbor Laboratory Press, Cold Spring Harbor, New York.

Hallowell E.M. 1999. *Connect: 12 Vital Ties That Open Your Heart, Lengthen Your Life, and Deepen Your Soul,* pp. 117–123. Pantheon Books, New York.

EXERCISES AND EXPERIMENTS

1 Mentors: Experiments with new behaviors

The purpose of this exercise is to help you to observe and evaluate how you interact with trainees. Below we provide a framework within which to record and assess your interactions. For those behaviors or interactions that were either ineffective or counterproductive, you will be asked later to experiment with alternative behaviors.

Record the incidents. During the next week or two, record two or three interactions with a trainee for which you felt the outcome to be unsatisfactory. For each incident, record

a. The incident or interaction

b. Your objective in the interaction

c. Your behavior

d. The effect of your behavior on the interaction or trainee

e. How your behavior affected the outcome or objective

Look for patterns. After you have accumulated several such incidents, examine them and ask yourself if you see a pattern. Does a specific person continue to come up? What about this person triggers your reaction? What are you experiencing during this reaction? Is a particular type of interaction more troublesome than others?

As an example, many mentors have difficulty giving negative feedback, reprimands, or evaluations to trainees and instead send confusing or ambiguous messages. If you have difficulty deciding whether interactions leading to unsatisfactory outcomes were influenced by your own feelings of discomfort or avoidance, refer to the tools for improving self-awareness presented in Chapter 2 for this part of the exercise. If you have difficulty discerning the effects of your behavior on others, refer to Chapter 6 and the exercises in that chapter on improving your observational skills.

Alternative behaviors. Propose alternative behaviors that might have led to a better outcome. List the possibilities you could have used (refer to the guidelines beginning on p. 121) and describe the anticipated outcome of the new behaviors. You may find it useful to record this information in the form of a chart, such as the one that we used for the case study earlier in the chapter.

Do the experiment. After you have listed several alternative behaviors and thought through their consequences, try one or more of your suggested alternatives in an interaction with a trainee during the following week. Determine whether the outcomes of the interactions involving the new behavior are better than those that occurred with your former behavior.

2 Trainees: Experiments with new behaviors

The purpose of this exercise is to help you to observe and evaluate how you interact with your mentor. Below we provide a framework within which to record and assess your interactions. For those behaviors or interactions that were either ineffective or counter-productive, you will be asked later to experiment with alternative behaviors.

Record the incidents. During the next several weeks, record two or three interactions with your mentor for which you felt the outcome to be unsatisfactory. For each incident, record

a. The incident or interaction

b. Your objective in the interaction

c. Your behavior

d. The effect of your behavior on the interaction or mentor

e. How your behavior affected the outcome or objective

If you have difficulty discerning the effects of your behavior on others, refer to Chapter 6 and the exercises in that chapter on improving your observational skills.

Look for patterns. After you have accumulated several such incidents, examine them and ask yourself if you can see a pattern. Does a type of interaction continue to come up? Does a specific behavior of your mentor trigger a strong response in you? What are you experiencing during this situation? Is a particular type of interaction more troublesome than others?

As an example, many trainees have difficulty hearing negative feedback or evaluations and instead hear confusing or ambiguous messages. If you have difficulty deciding whether interactions leading to unsatisfactory outcomes were influenced by your own feelings of discomfort or avoidance, refer to the tools for improving self-awareness presented in Chapter 2 for this part of the exercise.

Alternative behaviors. Propose alternative outcomes based on new behaviors. List the possible alternative behavior(s) that you could have used (refer to the guidelines beginning on p. 121) and describe the anticipated outcome of new behaviors. You may find it useful to record this information in the form of a chart, such as the one that we used for the case study earlier in the chapter.

Do the experiment. After you have listed several alternative behaviors and thought through their consequences, try one or more of your suggested alternatives in an interaction with your mentor during the following week. Determine whether the outcomes of the interactions involving the new behavior are better than those that occurred with your former behavior.

If you are a trainee, do the exercises at the end of Chapter 4. These are as relevant to your interaction with a mentor as they are with a supervisor in a traditional work setting.

CHAPTER 8

Science, Inc.: Make a Smooth Transition to Industry

A biotech executive recently said to me, "New scientists who join our company from academia are like deer frozen in car headlights. They don't know what's coming at them. They have no experience of working to time lines, reporting progress weekly, and having their work critiqued by every Tom, Dick, and Harry who happens to show up at a project meeting. It takes a couple of years for them to adjust."

Just as academic research labs present a spectrum of challenges unique to academia, so too do labs in the private sector. Scientists new to the private sector are rarely prepared for the different norms and expectations. The culture shock that scientists face when moving from academia is one of the most vexing problems of research enterprises in the for-profit sector. Your ability to sense and respond to changes in expectations will determine how effectively and quickly you adapt to new work environments.

Scientists often have a hard time accepting the loss of both autonomy and the ability to focus on "pure" research without regard for applicability, as well as the overriding concern with the bottom line. Scientists who tend to become emotionally bonded to their research projects can find themselves feeling manipulated in the private sector.

▸ **Corporate culture shock**
▸ **Mastering the art of scientific research in a corporate setting**
The only science that survives is profitable science
Understand your company's mission
Understand the market and economic factors that affect your company's business
Learn to give up projects
Do not expect to own your project
Understand the corporate evaluation and reward system
Learn to be a team member
Learn to get recognized for actions that support the group
The best solution or the first solution?
Get to the point
▸ **Silo thinking: The bane of large research and development organizations**
Silos in scientific organizations
Recognizing and avoiding silo thinking in science organizations
▸ **If you are a manager**
▸ **References**
▸ **Exercises and experiments**
1. Identifying silo thinking
2. Counteracting silo thinking

The following table summarizes several differences between academic and private sector research, with one or more personal or interpersonal consequences of each. The hard-nosed manager might be tempted to read through the list and respond, "Get over it!" Most scientists do get over it, but the loss of productivity during this transition is substantial and unnecessary.

Characteristic	Academia	Private sector	Consequence of difference
Autonomy	High. Scientific and administrative autonomy is the norm.	Low. The nature of for-profit research dictates that research be consistent with and lead to the advancement of corporate goals.	Scientists may feel manipulated or controlled, leading to resentment, low morale, and alienation.
Ability to follow own ideas	High/medium. Ideas can be followed within the scope of available time, funding, and manpower. Possibility exists for delving into new research domains.	Medium/low. Projects need to be conducted according to well-defined plans and time lines. Tolerance for exploration of peripheral areas is low.	Scientists may feel constrained and their creativity hampered.
Criteria for choosing a project	Typically chosen for scientific interest and importance to field. Choice may be constrained by interests of funding agencies.	Chosen for consistency with corporate mission.	Scientists may feel that their scientific input into project choice and focus is irrelevant.
Criteria for ending a project	Scientists find it hard to end projects in any setting. In academia, projects can persist for long periods of time, provided that funding is available. Often the only thing that can kill a project is the loss of a research grant, and even then, institutional funds are often available.	Projects can be killed for any one of a number of reasons, including change of company focus, loss of a corporate partnership, or failure to accomplish a predetermined set of objectives.	Scientists who are unfamiliar with the corporate mission or company politics may feel that project termination lacks justification and they were not given sufficient time to prove its worth.
Management of lab staff	Laissez-faire	Highly regulated	Need to learn new skills
Oversight of scientific work	Episodic and remote. Typically, oversight comes only after submission of a publication or grant application.	Constant. Scientists need to stay focused and deliver results.	Scientists may feel micromanaged.
Periodic evaluation of and input into personal performance	Virtually nonexistent	Depends on company, but typically ongoing and explicit.	Scientists need to be receptive to positive and negative feedback.
Need to work with multiple parts of organization	Typically of relevance only with respect to support services and administration.	Critical to success of complex projects. Need to understand and work productively with people who have distinctly different goals and viewpoints.	Scientists arrive unaware of the need for input from departments such as clinical, regulatory affairs, pharmacology, etc.
Stability of parent organization	Academic institutions are highly stable.	For-profit research organizations vary greatly in their stability. Start-ups may be especially unstable.	For the first time, job security may become a major factor in the scientist's professional life. Research outcomes may impact career and livelihood. Some may have difficulty maintaining objectivity.

CORPORATE CULTURE SHOCK

Scientists who join the private sector are often dismayed by the constraints to which they are subject. Seemingly simple decisions such as the choice of research tools and scientific approaches are often driven by the strategic focus of the company rather than an evaluation of the many possible alternatives. Scientists in academic settings exult in the discovery of an unexpected connection between their work and some other apparently unrelated phenomenon. Such discoveries are the stuff of new grant applications, funding, professional satisfaction, and prestige. In the private sector, such discoveries and insights must be sacrificed if they do not fall within the objectives of the organization. Finally, projects may end unexpectedly and for reasons that bench scientists find hard to understand. The following case study illustrates several of these issues.

> ▸▸ *Case Study: Terminated*
>
> Nano-Innovations was a three-year-old company with a staff of 30 scientists, most of whom had been recruited from academic labs. Nano had a collaborative project with ExploiTech who provided about $5 million a year in funding to develop a more efficient manufacturing process for the company's nanofibers. This project employed about 15 full-time Nano scientists. The agreement was renewable on an annual basis but ExploiTech had recently told Nano that they were unlikely to renew once the year ended. This news prematurely leaked to the scientists at work on the project, which still had six months to run under the terms of the present agreement. The scientists had difficulty understanding the termination decision, which had nothing to do with the quality of their work, but more with financial problems and management changes within ExploiTech. These scientists now faced six more months of work on a project that they knew would be terminated, and some uncertainty about their own jobs. Moreover, the scientists were frustrated that the project was ending because, with the encouragement of their CEO, they had been working long hours and were very close to making important improvements to ExpoiTech's manufacturing process. Some of the scientists resolved to never become this involved with a project again and to do only enough work to get by.

For a committed scientist, dropping a good idea or project can feel like giving up a child for adoption. Ending a favored project is something that even the most seasoned of research veterans finds difficult to do without much hand-wringing, procrastination, and last-minute reprieves. When I worked in academia, I walked into my laboratory on more than one weekend to find a dedicated scientist busily doing "one more experiment" for a project that I thought had been terminated weeks or months ago.

When a project is terminated for reasons that may have nothing to do with the promise or progress of the work, many scientists become disillusioned. Because scientists become intellectually bonded to the projects in which they are engaged, they often react by feeling frustrated and manipulated. Novice scientists, after being shuffled from one project to another, may avoid fully committing themselves to new projects to minimize disappointment. You do not need to be a scientist to recognize that such behavior is detrimental to any organization.

Such reactions are often attributed to individual personality and adjustment issues, or to immaturity. Although there may be some truth to these explanations, these reactions may also be a consequence of the poor preparation that scientists receive for working in the private sector and of the failure of managers to deal with the human consequences of scientific and business decisions. In short, both members and leaders of science organizations are deficient in skills that extend beyond the technical discipline of their specialty.

MASTERING THE ART OF SCIENTIFIC RESEARCH
IN A CORPORATE SETTING

If you are a scientist either planning to enter or who has recently entered the private sector, there is much you can and should do to minimize the chances that you will fall victim to corporate culture shock. The following sections help you to identify aspects of corporate scientific research that we know from experience cause consternation, confusion, and alienation in scientists. Your ability to recognize these problems and anticipate their effects on your attitudes and performance will improve your chances for a successful career in the private sector.

The only science that survives is profitable science

Like the previous case study illustrates, scientists need to be aware of the differences in the way that decisions are made in the private sector versus academia. Leaders of science and technology companies, especially those who are or were scientists themselves, have a penchant for motivating their scientific staff by extolling the elegance, power, and beauty of the company's science. They do this because they suspect that this is what the scientists want to hear. Do not be fooled by such oratory.

Be skeptical of companies that recruit research scientists with the promise of unfettered research opportunities. Very few companies live up to this promise, and many of those that try, especially new companies, change their tune after a few years. This is not to say that it is impossible to have satisfying and even exhilarating scientific experiences in the private sector—it is possible, and many do. But remember that the mission of your company is to create shareholder value and this may, at any time, lead to the curtailment or scaling back of your favorite research project. **The best way to insulate yourself from disappointment is to be informed about your company and the economic forces that impact it.**

Understand your company's mission

Understanding the mission of your company is a great way to keep abreast of which way the corporate wind will blow. But do not expect to learn anything from the mission statement displayed in the main lobby. Instead, watch and listen to what fellow employees do and say. Also, do not believe that your company's mission is unchangeable; after all, the life science research and discovery companies of yesterday are all scrambling to become the drug development companies of tomorrow—after spending a brief interval describing themselves as genomics companies.

Pay special attention to those programs that your company's senior management values most and to the goals on which they are focused. Unfortunately, it is not always easy to discern what senior executives are focused on. They may have a tendency to secrecy, especially when their focus may lead to mergers, layoffs, and downsizings that negatively affect employees.

Understand the market and economic factors that
affect your company's business

Look beyond the walls of your company to understand the world in which it exists. Scientists can be myopic in their view of their company and many are oblivious to the external factors that affect its fortunes. For those who transition from an academic setting, the notion that the destiny of their organization may depend on the availability of venture capital and changes in investors' perception of the technology's attractiveness (or "trendiness") may be an alien concept.

Arming yourself with information about the external forces that affect your company is the best way to limit feelings of helplessness and surprise when the unexpected occurs. Spend time learning about local market dynamics. During 2001, scientists who followed industry trade publication analyses knew that genomics companies focused exclusively on target identification were falling out of favor with the investment community. These people were not surprised when their companies announced that they would be downsizing their gene discovery efforts in favor of creating drug discovery operations. The more you understand about the market in which your company operates the better you are able to place decisions affecting your job in the proper context.

One middle manager in a biotechnology company related the following story:

"Our company started out as a combinatorial chemistry boutique. Our sole focus was to sell our compound libraries to biotech and pharmaceutical companies for use in high-throughput screening for drug discovery. My job was to lead a group developing screening assays. After years of explosive growth and high market valuations for companies such as ours, the investment community became skeptical. The focus of the biotech sector moved away from a fascination with clever technologies and more toward products, i.e., drugs. Because neither our company nor any company like ours had made much headway in actually making or helping to make an actual product, we took a big hit financially. We had to reorganize and we are now much more product oriented. Because I was watching this trend carefully, in early 2001, I began to volunteer for some of the more product-focused projects. My goals were to learn new skills associated with testing drugs in animals and eventually transition into the preclinical group if circumstances did not change. In fact, that's exactly what happened. Last year, we shut down most of our chemistry and screening operations, and added several new preclinical groups. I was made manager of one of the new groups but many from my old department were laid off."

If you are unsure of how to keep abreast of the industry trends that affect your company, ask the nearest vice president or, better yet, ask the CEO herself. Specifically, ask what to read on a weekly basis to follow industry news. This will get you noticed as someone who cares about the big picture (which immediately distinguishes you from most other scientists in the company) and points you to the same sources to which your superiors refer. Both are good ideas, provided that you are actually interested and pay attention to media resources.

Learn to give up projects

Although it may be difficult to give up a project into which you have invested much of yourself, the sooner you learn to get over it and move on, the happier you will be in the private sector. Companies with multiple projects and programs almost always track progress against predetermined goals or advancement criteria. Meeting these goals qualifies a project for continuation, while repeated failure likely targets the project for termination.

We have all heard stories of persistent scientists who argue doggedly and persuasively for the continuation of a project slated for termination. The typical outcome is that the project eventually results in a blockbuster drug or product. I have no doubt that such stories are true, but I also have no doubt that the vast majority of reprieved projects ultimately yield nothing. Those are the stories that you won't hear about.

There is no way to tell early in a project's life whether it will be a success or a failure. If you need to argue for the continuation of a doomed project, focus on scientific merit and supporting data. Be careful using arguments based on "scientific intuition" to justify the continuation of a program that has failed to meet its objectives.

Pay attention to any feelings of resentment toward management for terminating projects or making programmatic changes that affect you or your work. Ignoring or repressing these feelings may influence your behavior and attitudes in ways that have a negative impact on you or your relationship with management. It can be dangerous to harbor resentment when a project in which you had invested much psychological energy was cancelled, especially if you have poor self-awareness and self-control. These feelings can result in your becoming sarcastic or argumentative and may lead you to criticize management in ways that do you more harm than good. Refer to Chapter 2 for methods to improve your awareness of such feelings.

Managers may actually find it useful if you express feelings of resentment in the appropriate context in terms of the impact of decisions on yourself and how you feel. Telling your boss, "I was disappointed in how this project was terminated and I think that others were as well. Morale might be a bit low right now, and I thought that you'd like to know" may help her to recognize and address the impact of her decisions on employees. Telling your boss, "You people don't give a damn about how hard we worked on this project" may feel good at the moment but will not be very useful to either you or your boss.

Do not expect to own your project

In the private sector, individual scientists almost never own a project and often do not even own an experiment. Project ownership is hard to define because so many people are involved. The project may be managed by a program manager who has responsibility for coordinating the timing and execution of the various technical tasks that need to be accomplished. The tasks themselves may get done within specialized technical groups or "lines" managed by senior scientists or directors. Within these lines, individual scientists may be working on more than one project at any given time. In the worst cases, such arrangements can make scientists feel like cogs in a wheel. Scientists newly transplanted from academia, who are accustomed to having their own projects, often become concerned that their contribution to a project will become obscured. The result is that they may have difficulty motivating themselves to excel and to expend any extra effort. In fact, those who understand the corporate reward system learn that you do not need to own a project for your efforts to be noticed or appreciated.

Understand the corporate evaluation and reward system

Scientists in an academic setting have a pretty clear sense of how their performance is measured. They do experiments that either work or do not and participate in or lead projects that either succeed (and get funded or renewed) or fail. They submit papers to scientific journals that get either accepted or rejected. In each case there is a direct connection between the work of the scientist and the outcome or reward. As we show in Chapter 7, this connection fosters a powerful sense of ownership that is an important motivator for scientists in academia.

As we noted in the previous section, project ownership is harder to define in the private sector. This is not to say that scientists are not connected with or committed to their projects, because they typically are. What it says is that the outcome of a specific project will be less tightly coupled to the fate of scientists in the for-profit sector than in the academic sector. This is true not only for the organizational reasons outlined above, but also because in the private sector determining whether a project is a success or not can take an uncomfortably long time. For example, in the life sciences, the value of a drug target or a molecular drug candidate may not be known for 5–10 years after the early scientific research is completed. No one can be expected to wait 5–10 years for their bonus, nor should they. This is true in other technical disciplines as well.

Thus, instead of getting rewarded or evaluated for bottom line success or failure, you get judged on what might be called surrogate performance measures. You may not know whether the new receptor that you found will be useful as a therapeutic target, but you delivered it on time. You may not know whether your remote guidance system will help put a spacecraft on Mars, but it meets all of the criteria set by NASA. Management may not know whether the lead compounds that your team delivered will succeed in the clinic, but you produced 11 in the last year—one more than expected.

In the for-profit sector, scientists are typically evaluated annually. In well-organized companies, scientists and science managers have limited and well-defined goals and objectives for the year, sometimes for the quarter. These goals and objectives are typically divided into multiple categories, among which can be found specific project or scientific objectives. Because projects can last for years, each set of project-related objectives must be finite enough to be evaluated quarterly or annually. Thus, although it is conceivable that a scientist, or more likely a vice president, could have a goal that reads "develop a new drug against inflammation that acts via newprotein pathway," it is unlikely. The goal is more likely to read, "develop assays by which compounds binding to newprotein can be screened" or "validate newprotein as a target for inflammation by Q3." The more complex the project and the longer it takes, the more performance is measured by activities performed rather than results achieved.

In fact, the deeper within the organization one goes, the more removed a scientist's goals and objectives are from actual outcomes or accomplishments. I have seen team leaders struggle to create measurable and meaningful short-term goals and objectives for their scientists. Sometimes the goals include learning a new technique or attending a training course. In a scientific setting, in which important outcomes are the result of the combined efforts of an entire team or combinations of teams, it is difficult to assign meaningful scientific objectives to individuals. Despite this, scientists must maintain their focus on the goals that they set with their managers. As noted in Chapter 4, it is the job of the manager or leader to help you tie those goals into the big picture.

Learn to be a team member

In addition to getting evaluated by surrogate success criteria, you will also be appraised for your ability to work as part of a team and across or outside team boundaries. The importance of these evaluations in the private sector should not be underestimated. Unlike in academia, where you can sometimes find successful lab directors who have succeeded despite abysmal interpersonal skills, such instances are rare in the private sector. Companies cannot afford to place scientists who alienate others, or are argumentative or hostile, in leadership positions. Scientists in the private sector ignore this fact at their peril. I have seen highly skilled senior scientists lacking the most elemental interpersonal skills end up feeling marginalized and unsupported—without understanding the reasons, either because no one ever told them or because they were aware of the problem but incapable of doing anything about it. If you think you might fall in this category, avail yourself of every opportunity for feedback about your behavior from others.

On a psychological level, being a team member involves relinquishing some autonomy, being exposed to and influenced by viewpoints that may be radically different from yours, and making contributions that benefit the team's work rather than your own. If you do not have ownership and responsibility for your own project, you need to find other ways to stand out or distinguish yourself. This fundamental and natural need for recognition may cause difficulties for scientists in large organizations. The next section suggests ways to gain recognition as a member of a team.

Learn to get recognized for actions that support the group

Scientists have an expectation that individual knowledge, performance, and creativity will be recognized and rewarded (not only do scientists expect this, but so do their managers and supervisors). When I review a scientist's performance, I consider his contributions in meetings, the quality of his data, the way in which the data were presented, and the knowledge and insight he displays in the process. I also review how he interacts with, helps, or seeks help from others, along with a variety of other social factors. In many cases, I ask not so much what he accomplished, but rather how well he accomplished it.

Whereas teamwork is all about sharing, collaboration, and merged effort, the reward structure is all about getting noticed. Schein observes that this dichotomy can be damaging to team efforts: "We create tasks that are group tasks, but we leave the reward system, the control system, the accountability system, and the career system alone. If these other systems are built on individualistic assumptions, leaders should not be surprised to discover that teamwork is undermined and subverted." (Schein 1992, pp. 140–141.)

But this need not be the case. As we noted in Chapter 3, the opportunity for individual achievement and recognition and the focus on team objectives are not inimical to one another. If you are a savvy and adept player, you will learn to be successful in both contexts. You can be a team player and get noticed for your own individual achievements at the same time.

Some will never catch on, fearing that they will lose autonomy and power if they share their knowledge and insight with the team. Such scientists try to gain recognition and distinction based solely on their role as independent contributors. These employees become problems in team settings. The difficulty can become serious if their information and contributions are critical to the success of the project.

Because the need for recognition is strong, and because team rewards are seen more as fluff than substance, young scientists may hoard information or ideas and use them as currency to enhance their individual status. In one organization in which I worked, the running joke was that you could always tell when a certain scientist got an interesting experimental result because he would be seen literally running out of the lab into the CEO's office. It is not easy for scientists to shed behaviors and needs acquired during the formative years of their education and training.

If you donate your knowledge, data, and insights to the good of the project and the group, you will find that you will be viewed as both a team player and a valuable contributor in your own right. You do not need to demonstrate your knowledge in secret and you certainly should not do it in a way that embarrasses or demeans others.

The best solution or the first solution?

If your love of science is based on a search for truth and beauty in the natural world, you may have a rude awakening when you move to the private sector. Scientists delving into one scientific phenomenon invariably discover many other related phenomena that may be just as interesting, if not more so. As a basic researcher, you may want nothing more than the opportunity to follow your curiosity from one discovery to the next. However, you will rarely, if ever, have this luxury in the private sector.

Engineers seem to have an intuitive understanding of this reality. Engineering is all about working under a slew of constraints (mechanical, economic, temporal) to create something that works. Basic scientists often feel less comfortable with this mind-set. In many project meetings that I have attended, two implicit agendas were operating. The development group asked, "How can we take what we have now and make it work as a product?", and the research group wondered, "We really don't fully understand why this works. As a result, we can't guarantee that it will always work, and so we need to experiment a lot more before we're willing to turn it over to development."

This attitude is less common in established companies, especially in pharmaceutical firms, where multiple layers of oversight, tracking systems, professional project managers, and well-defined review procedures are in place. It is more likely to be found in small organizations with a high percentage of scientists newly arrived from academia.

The sociologist Herbert Simon noted that most complex problems are solved using an approach that he called "satisfycing" (Simon 1997a, pp. 118–120; Simon 1997b, pp. 28–30). Solving a problem by satisfycing means accepting the first solution that works, rather than seeking the best possible or optimal solution. Another way to put this is that "Organizations are happy to find a needle in a haystack, rather than searching for the sharpest needle in the haystack." (Allison 1971, p. 72.) Satisfycing offers the advantage of limiting the search for a solution to a subset of all possible solutions. Scientists in the private sector who attempt to manipulate all possible variables to ensure the best solution may be wasting their time. In most cases, the criteria for a good solution are not elegance or perfection, but how well the solution solves the problem within the constraints of available money, time, and manpower.

Because of their training and sometimes fastidious nature, scientists may be resistant to the notion of using the first good solution. Some will wish to find many possible solutions and choose among them. In many cases, this may be appropriate. For example, backup compounds for drugs are highly valued and appropriate. The art comes from knowing or judging when to stop. It is doubtful that there are rules to guide that process.

A company scientist needs to be comfortable making compromises between elegance and utility, full understanding and partial knowledge, and certainly and reasonable probability. Learning to choose between the best solution (which you may never find) and the first solution that works (which may not be the best) takes time. Many of the difficulties scientists experience in the private sector arise from the fact that the need to make such distinctions is rarely made explicit for them. Indeed, those at the corporate level are often unable to articulate the characteristics of a good solution. In any case, do not be misled by the fact that your group is called "discovery" or "research."

Get to the point

My friend Doug Kalish, a long-time veteran of the information technology world, recounted the following anecdote.

"In one of my companies, we hired many technical people right out of academia. I would assign them a project and ask that they come up with a solution that we could pass on to our development group and clients as soon as possible. I would then ask them to present their solution to the group. These presentations invariably contained dozens of slides that detailed every step, misstep, and approach that they took, and the conclusion was presented in the final one or two slides. I finally got so fed up that the next time they were scheduled for a project update I told them that I wanted to review their slides first. The next week, one showed me his presentation, which contained 50 slides. I looked at it for 30 seconds and said, "See this last slide, the one that you have as your conclusion? Start there. Throw out all of the other slides and tell us where we go from here."

When presenting their work, scientists and other technical professionals typically feel the need to recount every experiment and every dead end that they encountered before reaching their solution. We have all sat through seminars in which the speaker seemed compelled to show every experiment he did since graduate school.

We all want to get credit for how hard we worked and how clever we were to discover the right approach after slogging through all of the wrong approaches. But in the private sector, no one really cares. What is important is the end result and the data that support it. Your personal scientific journey is probably on no one's mind but your own. This is another difference between academia and the private sector about which no one will warn you.

This is not to say that there are never circumstances in which a recounting of each step in the experimental process is appropriate. In some instances, such as in a small technically focused team meeting, this may be expected. The key is to adapt your content to the audience and the circumstance. If you are updating a project for senior management, or presenting an overview to the board, lose the history and the personal saga. You will get more credit for being succinct than for being overly inclusive.

You may need to discover whether you have this affliction on your own, because not everyone is as forthcoming or helpful as my friend, Doug. Alternatively, if you are in the habit of overpresenting, you may end up annoying your audience to the point that someone feels compelled to inform you in exasperation. Pay close attention to your audience when making a presentation. Listen and watch, not just for their reaction to your scientific content, but for their reaction to you and how you come across. Your audience's body language and nonverbal cues may be your best early indicator of how you come across. If you sense that they are impatient, it is perfectly acceptable to pause and say, "I just want to check whether I am giving too much detail or background here. Is this level of detail okay, or would you prefer that I move right to the conclusion?" You may discover that you get more credit for being sensitive to your audience's level of interest than for the content of your presentation.

SILO THINKING: THE BANE OF LARGE RESEARCH AND DEVELOPMENT ORGANIZATIONS

The following two quotes are representative of what I hear from executives about communication among different parts of their organizations:

> "Our information systems group spent the better part of two years developing a new knowledge management system, but for reasons that I'll never fathom, they only discussed it with about three users in the entire company during that time. As a result, we're now six months postlaunch of the system and it is not being used because no one can figure it out."

> "Members of the discovery group start foaming at the mouth every time I send one of my marketing directors to sit in on one of their quarterly meetings. For their part, marketing thinks that the scientists are a bunch of clowns. As a result, one of our senior vice presidents has to serve as a go-between and this takes up half of his time."

The attitudes that lead to actions and reactions like those above are often referred to as "silo thinking." This is a metaphor for an attitude that creates a barrier between your group and others in the organization. The result is that each group functions in its own separate "silo" and communicates with other groups only as needed, or not at all. We have given silo thinking its own section in this chapter because its consequences to large science organizations are profound. Thankfully, application of improved self-awareness and improved interpersonal skills, especially in the realm of negotiation, can mitigate its negative consequences.

The principal reason for the creation of multifaceted science organizations is the need for contributions from those with widely different and complementary skills for creating a product or making a discovery. Silo thinking subverts this objective and results in absent or minimal communication among people or groups, delays, mistakes, and, in the worst cases, failure of products or projects. You do not need to understand the psychological origins of silo thinking to do something about it. In the following section, we show you how to recognize when you are succumbing to silo thinking and what to do about it.

Silos in scientific organizations

Perhaps the single most difficult conceptual barrier for scientists to cross in the private sector is that which separates basic or discovery research from the more applied or business-related functions. These include business development, clinical trials, marketing, regulatory affairs, and legal. The following table lists the various domains found in a typical life sciences company and one or more stereotypes (not entirely tongue-in-cheek) of that domain as it might be seen by those working outside of it.

Organizational domain	Example of cultural stereotype
Research	Impractical dreamers; no market sense. They fall in love with projects and never want to stop working on them.
Business development	Inundate research staff with half-baked proposals for in-licensing worthless programs. Conversely, company is unable to out-license its own programs.
Clinical research	Ossified clinicians whose notion of running clinical trials is stuck in the Neanderthal era.
Marketing	Slick salesmen who will try to sell anything to anyone, regardless of whether it works. They try to drive discovery based on market needs.
Regulatory affairs	Compulsive, humorless, rule-followers whose job is to place hurdles and impediments in the way of research and development personnel.
Senior management	Have no clue about what's going on in the labs.
Fiscal	These are the people who really make the decisions. They are the reason that no one can do anything.
Human resources	Try to train scientists to learn what they have no interest in learning. Force managers to spend time doing personnel evaluations that no one reads.

The cultural and personal gulfs that separate the domains listed in the table are legend and exist in companies of all sizes and types. Research scientists vigorously object when they feel that representatives of any of these groups attempt to influence the direction or focus of their work. From the researcher's perspective, the sole function of those outside research is to constrain them in their ability to do their job and shackle them with administrative requirements that make brilliant discoveries all but impossible. Conversely, business-related groups roll their eyes in horror when they talk about how strategic decisions are made by the research and discovery teams. In their view, these scientists are a bunch of ivory tower academics who, if left to their own devices, would develop products that have no intellectual property protection, require a 10,000-person 12-year clinical trial, and address a nonexistent market. These are, of course, extreme views and neither universally nor consistently held. But they are representative of views that I have heard expressed at one time or another in every company with which I have been associated. Inasmuch as these attitudes hamper communication with those outside of one's own group, they are damaging to the organization and its progress.

The following case study illustrates some of these difficulties. It is a composite case derived from my own first-hand experiences and reports from others who were themselves involved in the experiences.

▶▶ *Case Study: The Pet Project*

Alex is the program manager of a project at Mammoth Integrated Technologies to develop a new type of synthetic heart valve. This is a "pet" project of Shireen, the vice president for research and development, because she has a materials research background and knows the academic group that made the discoveries on which this program is based.

The chemistry group has synthesized modest amounts of several new polymers and the research team has been animal-testing valves made from these compounds. The results look promising, and Alex has petitioned the project review committee for authorization to move the program into the next phase of development, which would involve manufacturing more valves and getting the clinical groups involved to start planning for human testing if the animal studies progress.

This is the first meeting in which senior representatives from regulatory affairs, legal, and clinical trials are present. The CEO has also asked that a member from marketing attend.

The meeting opens with Shin, the lead scientist of the team, summarizing the progress of the program. Shin has been speaking for about ten minutes when the company's legal council, Sara, joins the meeting. After briefly listening to the discussion, she asks whether one of the valve manufacturing processes being used requires a license from the company that devised it. The head of manufacturing points out that they have used this method for over nine months, reporting progress monthly. Sara replies, "This is the first project meeting that I have been informed of and it's not my fault that no one involved me earlier in the process." Alex tries to intervene, but then Sara's cell phone rings and she steps out of the meeting and does not return.

Alex recalls that he had asked the vice present for research whether to include a legal representative at these meetings. The vice president cautioned him that this was not necessary at this early stage because "Sara has a way of bogging down project meetings in unnecessary details."

Shin continues with his presentation but is immediately interrupted by Oren, a physician and manager in the clinical trials group. Oren asks Shin to define the specific patient population for which he anticipates the new valve will be appropriate. Several members of the research group jump in with different ideas but for each idea, Ira, the vice president for marketing, weighs in with skepticism in an offhand and dismissive manner. By now, Alex and the other researchers are exchanging "I-told-you-so" looks. Shin sits down in frustration, feeling that the other groups were more interested in finding fault with the program than in hearing about his group's progress.

During this exchange, Alex resolves to himself that he will go to Shireen privately to argue for authorization to continue with the project as planned. He concludes that he will never get this group to agree and it would be best to simply forge ahead and allow the group to deal with results as they unfold.

Indeed, the next day, Alex meets with Shireen and convinces her that the clinical and marketing representatives are nit-pickers and cannot see that the project could lead to the most innovative product Mammoth has ever produced. Alex argues that the scientists are frustrated by the risk-averse stance of marketing, clinical, and legal and his group wants the chance to follow their scientific instincts. Given her personal interest in the project and her desire not to further frustrate the research team, Shireen agrees to allow the project to move forward.

We do not know whether the decision to move forward with the project was the right one. However, we do know that potentially valuable input from clinical, marketing, legal, and other groups was never properly integrated into the decision-making process. If the project moves forward without their participation, the team loses valuable input and may find itself with extra work to compensate for bad planning.

What happened here? Let us make a list of some of the events or circumstances that preceded this meeting.

1. Shireen is the scientific leader of the company but is intellectually bonded to the sci-

ence of the project, a danger sign that we noted earlier. As a result, she encourages Alex to move ahead without seeking valuable input from others in the organization.

2. Alex and the other researchers have a long history of feeling alienated from and frustrated by marketing and clinical representatives. The researchers tried to keep them out of the loop for as long as possible, so that when they did become involved, they felt marginalized.

3. Because the nonresearch groups could give little prior input, they felt "blindsided" by Shin's presentation and the fact that the project has continued for so long without them.

4. Shin bears the brunt of the resulting frustration during his presentation. He is naïve about clinical trials and marketing issues, and feels unfairly treated by those whom he views as outsiders with no understanding of the science.

5. Sara, the legal representative, has been actively excluded from any discussion of this project on the advice of Shireen. Shireen has, in effect, set Alex up for the confrontation that takes place in this meeting.

Although it may be unusual to find all of these problems in a single project, most projects manifest one or more of them at some point. Some will read the above account and focus on the various ways in which the case represents an example of failure to adhere to good project management principles. But project management, like science and technology, can be known and understood but poorly implemented.

Successful project management requires that the participants engage in face-to-face interactions, conversations, and meetings with those outside of their own areas of expertise. The research team needs input from legal to ensure that they are free to use the required methods and techniques. Conversely, legal needs input from research to plan their strategy for protecting the intellectual property that the project generates. The research team also needs to communicate with the clinical team so that they can decide on the appropriate patient population for their new valves. This decision is of great importance, since it will also influence the kinds of animal models the research team uses for testing. The clinicians need input from the researchers so that they can make the most appropriate recommendation and decide on other possible clinical indications for the new valves. Similar interactions must take place with marketing, regulatory affairs, manufacturing, engineering, and other groups within the company.

Each of these interactions has the potential to create discomfort for the participants. None of them fully understands the others' expertise, and each may use language or terminology with which the other is unfamiliar. A research scientist may not really understand the distinction between a Phase 2a and Phase 2b clinical trial, and a clinician may not appreciate the subtleties of why the valve that he is testing is unstable in acidic media. Also, seemingly universal concepts such as time, which participants take for granted, may be viewed differently by each. In *Organizational Culture and Leadership*, Schein reviews studies showing that those in sales and marketing view time as having a very different (typically, orders-of-magnitude shorter) horizon than do scientists in research and development. The result is that when one group refers to completing a project "on time," the other group may have its own notion of what that means and neither recognizes the misunderstanding (Schein 1992, pp. 110–112).

As a result, conversations among members of different groups may be full of misunderstanding and false assumptions. If the participants have a preexisting skepticism about the competence, motivation, or importance of one another, the likelihood of a productive discussion diminishes further. These problems can be attributed to the difficulty of individuals to deal with and function in ambiguous or stressful situations. Such difficulties are inevitable and even natural in large organizations. Your objective as a member is not to eliminate these difficult interactions, which is impossible, but rather to work though them productively.

Here is our analysis of the causes of the problems manifested in this case study.

1. As vice president for research and development, Shireen owns significant responsibility for the difficulties. She must become aware of her tendency to champion and protect projects in which she has a personal interest. Knowing this will sensitize her to the likelihood that she will yield to the research team's desire to be isolated from outside input. Shireen must also become aware of her attitudes toward members of other groups in the organization, including legal. It is almost certain that she is passing her views and attitudes to her staff, thereby perpetuating her own biases and further solidifying the silo thinking of all involved. Shireen should address specific issues having to do with Sara's behavior at project meetings directly with Sara. Her reluctance to do this may stem from her inability to have what she expects will be a difficult conversation with Sara (see Chapter 6 on dealing with peers).

2. As project manager, Alex has a responsibility to build alliances with the various support and development groups and set an example for the members of the research project team. The ability to negotiate during uncomfortable or contentious situations is perhaps the single most important attribute of a successful project manager. Alex's inability to either anticipate or deal with the strong reactions during the project meeting suggests that he needs to work on acquiring these skills.

3. Both Alex and Shireen failed in their responsibility to educate the research team about the roles of the development and business groups. Moreover, they both encouraged the belief that these groups were mostly a pain in the neck. As a result, the research scientists' skeptical and dismissive attitude is readily apparent during this and previous meetings, and exacerbates the atmosphere of mistrust and antagonism among company domains.

4. Despite the fact that he is a junior member of the team, Shin had a responsibility to understand the roles of the clinical and marketing groups in his organization. Had he been more aware of these, he might have anticipated the comments of these groups' representatives or, at the very least, taken them in stride.

Had even one of the participants been aware of what was happening from the perspective of silo thinking, and had they taken steps to bridge the differences among the various groups involved, the outcome of the meeting could have been quite different.

The proximate causes for the way the final decision was made are a combination of organizational, leadership, and individual behaviors that can and should be addressed by companies seeking to improve performance. The following section describes an approach to addressing silo-related problems.

Recognizing and avoiding silo thinking
in science organizations

Recognizing silo thinking really *is not* rocket science. Ask yourself the following questions:

- Do you or others in your group routinely criticize members of other groups in your organization?

- Do you notice a company or group mythology about misguided or inept behavior on the part of members in specific groups?

- Do you try to avoid having certain members from other groups or departments at meetings or research presentations?

- Should or must parts of your organization have input into what you do, but you have never understood their precise role, or you doubt their value?

- Do you find yourself regularly being dismissive or skeptical when someone from one of these other groups offers input?

- Do you find yourself routinely feeling annoyed or argumentative when you speak with particular people from other groups?

   If you answered yes to one or more of these questions, it is likely that your work and your organization's progress are being hindered.

Learn to identify silo thinking when it happens

The first step to overcoming silo thinking is to recognize it. Pay attention to your interactions with those from other groups in your organization. If you have difficulty concentrating on doing this during the interaction, make a point of debriefing yourself afterward. This is especially important when you have what feels like an uncomfortable or confrontational interaction. Refer to the first exercise at the end of this chapter.

Evaluate the consequences of silo thinking

It is not necessary to like everyone in your organization. But you do need to work with them. When professional interactions are impeded by your personal feelings, the consequences may be detrimental to your work, the advancement of your project, and the organization as a whole. You have a responsibility to anticipate and overcome such behaviors; Chapter 2 shows how this can be done.

Work to build bridges to other parts of your organization

You also have a responsibility to build bridges to other parts of your organization, which is also an effective way to distinguish yourself. It shows that you appreciate the need for input from and collaboration with those outside of your own area of expertise. In addition, it indicates that you are not afraid to enter into discussions with those who may not fully understand what you or your group does. These are qualities that well-managed organizations recognize and reward. Refer to the exercises at the end of the chapter for additional help in developing alternative behaviors toward those who you victimize by silo thinking.

It will not be possible to change an ingrained pattern of antagonistic interactions overnight. Building bridges involves taking small steps that change behavior and interaction patterns over time, and in a way that allows group members to become acclimated to the changes. Simple ways to start this process include

- Asking a member of another group for information about the group's function, responsibilities, and organization.

- Actively seeking the point of view of a member of another group rather than waiting for them to offer it.

- Choosing one person within another group as a contact and using him to better understand the perspective of others in the group.

The exercises at the end of the chapter list some ways in which you can modify your own behavior to begin mitigating the effects of existing silos. These exercises are based on the notion that as you reach out to other groups in a nonconfrontational, supportive, and open manner, they will eventually respond in kind.

Learn to negotiate in a way that fosters a fair and equitable outcome

Many interactions across organizational domains involve reaching agreements with others who seem to have very different viewpoints and agendas from your own. Circumventing the process by going to a higher authority, as Alex did in the previous case study, almost always results in increased antagonism and typically only postpones the inevitable showdown. Learn to resolve such issues on your own by identifying and focusing on your and the other party's underlying interests (as opposed to positions). By negotiating with equanimity in the face of hostility and stubbornness you can distinguish yourself from those who either give up or become hostile.

It would be naïve to believe that improved interpersonal or negotiation skills will break down every barrier to communication in science organizations. Some in other departments may resist your efforts to build bridges. Others may be unwilling to listen because they sincerely believe that they have no need for your input. Your organization may be led by someone with a pathological personality who intentionally sets people against one another. Fortunately, these are truly the exception rather than the rule. Most organizational members like their work and want to do a good job, but have just as much difficulty crossing cultural divides as you. Learning to build bridges across those divides is one of the most important skills that you can learn.

IF YOU ARE A MANAGER

A manager must take responsibility for educating new recruits from academia in the mores of the private sector. Most science managers have never had the differences explained to them, and therefore may not fully understand them and their consequences. Scientists need to understand how decisions are made, what scientific and economic factors affect the decision to create or terminate projects, and how rewards and promotions are allocated. Review the sections in Mastering the art of scientific research in a corporate setting, starting on page 132, to ensure that new recruits from academia understand the various ways in which science in the for-profit sector, and in your organization in particular, differs from academia.

Help your staff to avoid silo thinking. Pay attention to your own attitudes toward members of other divisions or groups. If you feel negatively about them, you will likely convey this to your employees unless you maintain an awareness of these attitudes.

In summary, multiple opportunities exist for projects to succeed, fail, or head down unexpected, sometimes beneficial, paths. In many cases, it is the quality of the interactions among the participants that determines the outcome. The subtle observations of scientists involved with the performance of assays, manufacture of devices, or testing of new technologies can be communicated to others—it all depends on the quality of the interactions among members of the organization. A willingness to share hunches, intuitions, and "crazy ideas" cannot be mandated by the formation of interdisciplinary project teams. The ability of individuals to communicate with others outside of their own domain of expertise will determine the way and facility with which these interdomain issues get discussed, explored, or resolved. The effectiveness of such communications can be influenced to some degree by the type of "organizational culture" found in a company. But it is most strongly influenced by the ability and skill of the individuals involved to engage in productive interactions.

REFERENCES

Allison G.T. 1971. *Essence of Decision: Explaining the Cuban Missile Crisis*, p. 72. Little, Brown, Boston, Massachusetts.

Schein E.H. 1992. *Organizational Culture and Leadership*. Jossey-Bass, San Francisco, California.

Simon H.A. 1997a. *Administrative Behavior: A Study of Decision-making Processes in Administrative Organizations*, 4th edition, pp. 118-120. The Free Press, New York.

Simon H.A. 1997b. *The Sciences of the Artificial*, pp. 28–30; pp. 119–121. MIT Press, Cambridge, Massachusetts.

EXERCISES AND EXPERIMENTS

1 Identifying silo thinking

You probably already know who or which groups within your organization that you view with antagonism. Perhaps their values and focus are different from yours. One common divide within companies is between "research" and "development." An important step to becoming a skilled manager is identifying those whom you view as outside of your silo and examining the rationale and consequences of viewing them in this way. Who do you criticize? Who annoys you? The next time you react in one of these ways, try listening instead of judging. Put yourself in their place. Understand their mission. See yourself as they see you.

- Do you or others in your group routinely criticize members of other groups in your organization? If so, list the individuals and/or their groups and indicate how their function is connected to that of your group.

- Do you notice a company or group mythology about ineptness or incompetencies on the part of the members of specific groups? List one for each group.

- Do you try to avoid having certain members from other groups or departments at meetings or research presentations? Name them, their groups, and your reason for avoiding them.

- Should or must parts of your organization have an input into what you do, but you have never understood their precise role? List them and describe what you do not understand about their function.

- For one or more of the groups that you identified above, list three negative characteristics. For these same groups, list three negative characteristics that you suspect they might attribute to your group.

- For each of the groups, list one or two consequences to your organization of poor or strained communications among your group and the others.

2 Counteracting silo thinking

Open a line of communication. Chose an individual in each of one or more of the groups that you identified above. Ask this person two or three neutral information-gathering questions regarding their groups' functions. Listen to their answers with openness and curiosity.

Be supportive when you might normally be hostile. Chose one of the groups from the list above. The next time that you find yourself together in a meeting, say something supportive to someone from that group with whom you might normally disagree or ignore. Refer to Chapter 5 for guidance. You can always find something with which to agree even if you disagree on the major point. Continue until you get a reciprocal response from that person or from someone in that group. Note whether your perception of that person or group changes over time and whether you notice an improvement in group interactions in general.

CHAPTER 9

Shape the Future of
Science and Technology

Recently, the editors of a scientific magazine with a large national distribution
asked me to write a brief response to
two questions: How should senior scientists
and principal investigators manage their labs
to avoid conflict between lab members? and
What is the best way to handle strong personalities? When I told them that I would be
happy to write an article on these topics,
they said that they were only looking for a
few bullet points to place in the magazine. I
respectfully declined and advised them to
await my book.

Learning to be more self-aware and work productively with others is no simple task
and can hardly be captured in a few bullet points. Even though we have introduced tools
and concepts to help you along this path, we do not for a moment believe that it is an
easy one. First, changing how you see yourself and interact with others may require you
to alter lifelong and deeply ingrained traits and habits. Second, science and technology
communities and the organizations and members within them are resistant to change.
Yet change can happen, even if only by one person at a time.

PUTTING IT ALL TOGETHER

We present one final case study that illustrates the possibilities for change. The case is
presented in more detail than our previous cases because we wish to depict the complexity of real-life situations. It is based on a true story, but has been significantly altered.

As you read through the case, note the problems or potential problems that you see.
You may find it helpful to categorize the difficulties you encounter in terms of the themes
of the previous chapters. Do you find issues related to self-awareness, self-control, peer relations, mentoring, group management, negotiation, etc.? In addition, write down how you
might have handled those specific problems differently from the participants in the case.

At the end of the case, we present alternative hypothetical outcomes to illustrate how
several aspects of the case could have been handled differently by one or more of the protagonists.

▸▸ *Case Study: Rewriting History*

Susan was a senior investigator in the structural biology unit of USA Pharma. She worked in a group loosely organized under Ed, the head of structural biology. Her project focused on the three-dimensional structure of a growth-factor receptor called PAN that had been implicated in pancreatic cancer. Susan was an ace at X-ray crystallography and was recognized as an expert in receptor structure even before she joined USA. But Susan had problems. Ed noticed from day one that she was secretive and reluctant to share her insights and data with others. Ed was dismayed at this, but took a rather laissez-faire attitude. Her work was superb and he was reluctant to make waves for fear of alienating or losing her. Also, every time he broached the topic of openness with Susan, she became defensive and angry. Ed and most of the other scientists at USA felt at a loss in these situations and decided to simply ignore the problem and hope for the best. After all, Susan was producing good results, so better leave well enough alone. He was reluctant to bring the problem to human resources because he did not have faith in them and thought that they might just make matters worse.

Athan, another investigator at USA who was several years senior to Susan, had been working on PAN as well. In fact, he was taken by surprise when he learned from Ed that Susan would be working on PAN, but he was reluctant to say anything about it. Ed had managed another project in a similar way once before, but when Athan queried his decision, he denied having any ulterior motive beyond moving the science forward, and he cut the discussion short by saying that members of the lab had to resolve such matters on their own.

Athan's relations with Susan had been tense. Shortly after Susan arrived at USA, she noted that Athan used a sample preparation technique that she believed was totally inappropriate for PAN-type proteins. She told him this in such a condescending manner—referring to him as "an amateur"—that Athan was both hurt and angry for days.

Athan was uncomfortable with Susan's style: She was direct, argumentative, and to the point, whereas he was much more circumspect and reserved. As a result, Athan and Susan rarely spoke to one another. Although they did not get on too well, they were well aware of one another's data. At one point, after seeing some of Susan's especially exciting data, and despite their earlier clashes, Athan suggested that he and Susan collaborate. Susan turned red in the face and practically shouted that she did not need him to help her interpret her own data.

When Ed got wind of this confrontation, he called them both into his office. He said nothing about their interpersonal clashes. Instead, he told them that to avoid overlap in their work, they must each confine their efforts to different aspects of the PAN problem, which he delineated for them. They agreed, but relations between them never improved.

In October, two investigators in the cancer biology division of USA, Mohan and Ralph, told Athan that they believed PAN was related to another receptor called MBC that they had implicated in certain malignant brain tumors. In fact, they had already developed lead compounds that inhibited the activity of MBC. Unfortunately, these compounds had no effect on tumors in animal models, and Mohan and Ralph were being pressured by Ruth, their supervisor, to abandon the project. They argued against this and asked if they could determine whether any of their lead compounds might be effective against pancreatic cancer before they put the project aside. Ruth would not authorize synthesis of more test compounds, or new animal studies, without more definitive data to support the similarities between the receptors. So Mohan and Ralph wanted to take a closer look at the PAN structural data.

Mohan and Ralph had previously met with Susan on two occasions during the past 12–18 months. Mohan in particular was put off by her abruptness and abrasiveness. After each meeting, Mohan came away feeling that a give-and-take discussion would be impossible. More than once, Mohan felt insulted by Susan's comments, and became angry in return, exacerbating an already difficult situation. Eventually, Mohan decided that dealing with Susan was just too uncomfortable and, because of other commitments, temporarily shelved the PAN problem.

For her part, Susan felt that Mohan was a naïve dilettante when it came to receptor structure. Since she had never felt the need to suffer fools gladly, she was dismissive of him. Sharing data was not an issue, but having to deal with someone who tried to tell you your own business was quite another.

Despite the discomfort, at the second of their meetings, Mohan and Ralph were able to ask Susan about the similarity of the PAN receptor to MBC. Susan replied that she had suspected for well over a year that the receptors were related, but that this "didn't mean a damn thing" and she needed more data to establish this definitively.

Now, 18 months later, during a casual conversation about the PAN project, Athan showed Mohan and Ralph both his own data as well as some of Susan's new data. He also showed them Susan's slides that contained her close-to-final structural model for PAN. After a brief and excited analysis by Ralph, they all concluded without question that PAN was of the same class of receptor as MBC.

In a burst of enthusiasm, they received Ruth's approval to synthesize more of their lead compounds and sent them off to one of their academic collaborators, who was an expert in pancreatic cancer. Within two months, the results showed that the compounds were highly effective at slowing or stopping the growth of this cancer in mice. Based on this information, USA decided to move the candidate drugs into human testing, and in a short time they obtained promising positive results.

Athan felt very uncomfortable about this whole process. He knew that if Mohan, Ralph, and Susan had been able to forge a collaboration sooner, or at least had had a discussion that did not result in them all going away insulted or angry, a therapy for this deadly cancer might have been available to patients at least two years sooner than it would be otherwise. Moreover, he also knew that the financial implications for his company were huge. During the next six months, Mohan and Ralph were widely credited for their roles in discovering the PAN therapy and Athan was given credit for assisting them. Susan's contribution was downplayed and Mohan, Ralph, and Athan did not go out of their way to correct this impression. No one seemed to remember that Susan herself had seen the similarity between PAN and MBC before anyone else. Within the year, Susan left USA for another job.

This vignette is based on a true story. In the modified version, the two-year delay in the development of the pancreatic cancer therapy would have been unfortunate for USA and possibly fatal for many cancer patients. Could it have turned out otherwise? In many ways, the outcome to this story could have been different. Below we examine several alternative scenarios, focusing first on Ed's role. Let us take Ed's perspective, and pick up the story just before Mohan and Ralph approach Athan.

▸▸ *Case Study: Rewriting History (revised version)*

For some time, Ed was concerned about both Susan's protectiveness of her data and her hostility. He felt that it was detrimental to the group and would certainly not benefit Susan's career, either at USA or else-where. After some thought, Ed realized that his own reluctance to deal with Susan was a big part of the problem. He could not handle conflict. His tendency was to avoid angry people, especially women, at all costs. He realized that in the past, he had let hostilities in the lab simmer and hoped that they would just disappear. He had behaved this way with Susan for two years and concluded that he was doing a disservice both to Susan and the company.

Ed also suspected that he knew the reason for Susan's paranoia about her data. When he hired her, he had deliberately assigned her to the same project on which Athan had already been working for two years. He had tried this approach before with the idea that either the synergy or the competition would accelerate the project. In this case, some benefit resulted, but the tense atmosphere created by the situation had made the whole department uncomfortable. Thus, Ed detected multiple problems, including his poor judgment in assigning projects, Susan's abrasive manner in the lab, and his own unwillingness to intervene.

For her part, Susan believed that others in the group—mostly men—were given advantages that she was denied. She felt that she had to prove herself all over again every time she joined a new company. The only way that she could ever get any respect at USA, she believed, was to solve the PAN problem and prove her capability.

After much thought and preparation Ed called Susan into his office and had the following conversation with her.

"Susan, I know how hard you've worked on the PAN project and I believe I know how important the project is to you. Although I can empathize with you, I'm frustrated by the divisive atmosphere in the department. I know that you get along with many people, but are you aware that you have a tendency to be confrontational with others? This behavior is not in the best interest of the company or you. Frankly, I even had difficulty deciding to discuss this with you because I was afraid that you would get angry. It's not easy for me to talk to people who get angry so I've avoided being honest with you. I'm sorry that I haven't discussed this with you before, but I'm doing it now because you're a very important member of the department and we need your talent and skills.

I suspect that you were frustrated because of the way your project was assigned. Is that true? If so, I take responsibility for that. Moreover, as head of the department, I feel that I haven't been honest with you and I haven't given you the opportunity to grow professionally. I'd like to correct that now. I need to give you some honest feedback about how your anger impacts our working relationship. Then I'd like to work with you to help you change that behavior. If you're frustrated with the way in which you have been treated, we can deal with that to see if we can fix it. I'd also like to help you deal with Athan, Mohan, and Ralph in a way that allows you to express your scientific views clearly and without anger. Do you agree that there is a problem?

I know that you're a dedicated scientist and that getting the science right is your primary motivation. I think that I can help you to deal with and control your anger in certain situations so that you get your scientific points across without others feeling that you are disrespectful of them.

This is not to say that you don't have reason to be angry at times. You may have a very good reason and we can and will discuss the substance of those issues. But right now, the priority is learning how to best express your concerns in ways that give you the greatest opportunity to be heard and have a positive impact on the project."

Ed concluded, "Look, we're all human. If we're angry or sarcastic, we're bound to generate a negative response even when what we say is demonstrably and scientifically true. People may hear what we say, but they may not take it in because they're reacting to their own feelings of being insulted, hurt, or angry. It doesn't have to be that way.

Finally, having said this, I need to tell you that you do need to improve your interactions with others in the company. I am convinced that the difficulties you have in working productively with others slow the progress of our projects, and I can't let that continue. I need your commitment that you will work to improve your professional relationships, and we need to set some specific goals for improvement. I'm willing to work with you on this because I value your contributions. Is this something that you're willing to work on with me?"

We will never know what the effect of such a conversation might have been because in the real case, it never took place. But we can imagine. We can imagine that Ed's non-confrontational discussion with Susan, his honesty about his own feelings, and his responsibility as a mentor might have had some impact on Susan. After some time, with guidance and feedback, she might have learned to control her anger and focus her energies on building alliances with some and working collaboratively with others.

Let us analyze what Ed said to Susan in the second version and see how he used many of the approaches we have introduced in the book.

"Susan I know how hard that you've worked on the PAN project and I believe I know how important this project is to you. Although I can empathize with you, I'm frustrated by the divisive atmosphere in the department. I know that you get along with many people, but are you aware that you have a tendency to be confrontational with others? This behavior isn't in the best interest of the company or you. Frankly, I even had difficulty deciding to discuss this with you because I was afraid you would get angry. It's not easy for me to talk to people who get angry so I've avoided being honest with you. I'm sorry that I have not discussed this with you before, but I'm doing it now because you're a very important member of the department and we need your talent and skills.

I suspect that you were frustrated because of the way your project was assigned. Is that true? If so, I take responsibility for that. Moreover, as head of the department, I feel that I haven't been honest with you and I haven't given you the opportunity to grow professionally. I'd like to correct that now. I need to give you some honest feedback about how your anger impacts our working relationship. Then I'd like to work with you to help you change that behavior. If you're frustrated with the way in which you have been treated, we can deal with that to see if we can fix it. I'd also like to help you deal with Athan, Mohan, and Ralph in a way that allows you to express your scientific views clearly and without anger. Do you agree that there's a problem?

I know that you're a dedicated scientist and getting the science right is your primary motivation. I think that I can help you to deal with and control your anger in certain situations so that you get your scientific points across without others feeling that you are disrespectful of them.

Principles used

Prepare. Ed has prepared for this discussion by reviewing his interests. He has concluded that his strongest interest is in maintaining the productivity of his group. Therefore, he needs to keep Susan and find a way to improve her working relationships with others.

Empathize to defuse or forestall anger. Ed starts by empathizing with Susan, recognizing the importance of her work as well as her frustration.

Ask questions. Ed ask questions to ensure that Susan sees the problem and understands what he is saying.

Be attentive to reactions. Ed carefully watches Susan's facial expressions and body language to be sure that she is willing to discuss her behavior. This is a critical point in the interaction: Ed must ascertain whether Susan is both willing and capable of discussing her own behavior and prepared to take responsibility for it. If at this point she becomes angry and denies the problem, or insists that the problem is entirely due to how others treat her, Ed will need to tread carefully and either terminate the discussion or jump right to the last paragraph to discuss his expectations.

Develop self-awareness. Ed has thought through his interactions with employees and peers and knows that he tends to avoid conflict. This alerts him to the fact that he might have done the same with Susan.

Create an atmosphere of trust. Ed accepts responsibility for his role in the problem, further enhancing the chances for an honest and nonconfrontational discussion.

Accept responsibility. By taking responsibility for his role in the situation, Ed creates an atmosphere of trust.

Focus on the problem, not the person. Ed frames the problem in terms of its consequences to the company. He avoids attacking Susan.

This is not to say that you don't have reason to be angry at times. You may have a very good reason and we can and will discuss the substance of those issues. But right now, the priority is learning to best express your concerns in ways that give you the greatest opportunity to be heard and have a positive impact on the project."

Ed concluded, "Look, we're all human. If we're angry or sarcastic, we're bound to generate a negative response even when what we say is demonstrably and scientifically true. People may hear what we say, but they may not take it in because they're reacting to their own feelings of being insulted, hurt, or angry. It doesn't have to be that way.

Finally, having said this, I need to tell you that you do need to improve your interactions with others in the company. I'm convinced that the difficulties you have in working productively with others slow the progress of our projects, and I can't let that continue. I need your commitment that you will work to improve your professional relationships, and we need to set some specific goals for improvement. I'm willing to work with you on this because I value your contributions. Is this something that you're willing to work on with me?"

Principles used

Focus on the problem, not the person. Rather than being accusatory, Ed focuses on how Susan's behavior affects her work and her professional relationships. He maintains the focus on her behavior, not her personality, and offers support.

Focus on the problem, not the person. The problem is decreased productivity due to poor communication and interactions within the company.

Notice body language. Ed watches Susan's body language and notices that she begins to nod almost imperceptibly as he speaks. Picking up on this cue, he gives her an opportunity to speak.

Focus on the problem, not the person. Ed makes it clear that his goal is to advance the company's programs and science, an aim with which Susan can identify.

Identify underlying interests. Ed maintains his focus on his interest in getting the company's work done and on Susan's interest in having an impact on the company.

Get a commitment. Ed concludes by making it clear that change is needed and why. He reiterates his underlying interests and asks for Susan's commitment to work with him. This last statement ensures that Susan buys into a commitment to change.

Assure. Ed assures Susan of her value and his willingness to help her.

An important element in this interaction is Ed's observations of how Susan responds to his analysis of the problem and his suggestions for improvement. As we show later in this chapter, Ed's assessment of whether Susan is willing and able to engage in this type of discussion, and whether she accepts responsibility for her own behavior, is key to the success of his attempted intervention. If early in the discussion Susan becomes indignant, denies any responsibility for the situation, and blames the strained relationships on others, Ed may conclude that this approach will not be successful. In that case, he has to decide to either let things continue or find a solution that alleviates the problem but does not require that Susan have insight into herself. One possibility might be to appoint another scientist to serve as liaison between Susan and the others. However, if what is needed is Susan's insights, personal knowledge, and active participation in a joint project, there may not be an easy solution.

In the above scenario, we focused on how Ed could have used his role as mentor to improve interactions and collaboration between Susan and others in the company. Of course, the situation could have been improved in many other ways. Let us list a few of them.

1. Susan might have read this book and started taking notice of her own behavior during interactions with her peers. She might have concluded on her own that some people, Athan and Mohan especially, pushed her hot buttons. She might have discovered that she was especially sensitive to having her work questioned in the kind of glib manner that Athan and Mohan seemed to favor. She felt that they were arrogant and dismissive of her, and in some cases they actually were. Had Susan come to these conclusions, she might have decided to ignore their behavior and to observe, rather than react, to the feelings engendered by those behaviors. Getting some distance on her feelings might have given her the opportunity to choose how to behave rather than simply act in response to her feelings of anger or annoyance.

   This is not to suggest that Susan should have stood quietly by if she was insulted, mistreated, or ignored. Rather, our suggestions might have allowed her to deal with these problems and satisfy her underlying interests (doing good work and advancing the project) without resorting to either withdrawal or hostility.

2. Had Athan read the section on conflict avoidance in Chapter 4, he might have noticed that Ed was conflict avoidant and this was harming the project. Athan might have concluded that Ed was either blissfully unaware of or consciously ignoring the difficulties that he and Susan were having because he did not know what to do about them. If Athan had been trained in conflict resolution, he might have felt comfortable approaching Ed and outlining the difficulties as he saw them. Taking negotiation cues from Chapter 3, Athan might have appealed to Ed's underlying interests in ensuring that his projects progressed, getting Ed to focus on the several problems needing his attention.

3. Had Mohan read this book, he might have been successful at developing a collaborative relationship with Susan. By ignoring or deflecting her insults (Chapter 8) and using techniques to defuse and neutralize her anger (Chapter 4), he might have been able to overlook Susan's tough exterior and maintain his focus on developing a relationship that would have furthered the interests of the company. By framing discussions with Susan in the context of the problem (the company's need for diffusion of information across departments) rather than the person (accusing Susan of being obstructionist or hostile) as described in Chapter 4, Mohan might have been able to appeal to Susan's highly developed professional standards.

These brief alternative analyses, in which we focus on how changes in individual behavior might have resulted in alternative outcomes, should not be taken to imply that any one of the individuals was either at fault or solely responsible for the outcome. Although it may have been possible for any one of them to individually initiate a resolution of the problem, they are all jointly responsible for its genesis.

After reading this case, it is tempting to pinpoint Susan as the problem. But this is a mistake. Closer analysis reveals many contributing elements, from Ed's initially assigning Susan to work on the same project as Athan without his knowledge, to Ed's avoidance of the resulting problems, to Mohan's inability to deal with Susan productively. Susan's hostility is, of course, a major factor in this case, but it does not stand in isolation. It is easy to blame individuals for group problems because individual behavior is so easy to see. In extreme cases, the behavior of one individual may seem so aberrant that it may seem logical to identify it as the cause of a group problem. It is harder to see the web of relationships and interactions that inevitably underlie all problems in groups. In our experience, focusing attention on one group member as the singular cause of a group problem is almost always a mistake.

Focusing on an individual also diverts attention from the group's underlying interests. In this case, the interests of USA and its staff are improving interactions, sharing information, and collaboration among its scientists. As we have seen in the alternative scenarios above, these interests might have been addressed in various ways that would have avoided using blame or scapegoating.

Because this case takes place in a company responsible for a lifesaving and financially lucrative product, the benefits of any of the interventions suggested above to Susan personally, the company, and sick patients could have been huge. If the two-year delay in the delivery of the product did not occur, it might have saved the lives of many patients. Susan might have received just credit for her role in the discovery, rather than being marginalized and ignored. (*Note to the reader:* The scientific details of this case are purely fictional. The PAN receptor does not exist, and unfortunately, as of this writing, nor does a treatment for pancreatic cancer.)

But even if this case had taken place in an academic setting, an alternative outcome could have been equally important, at the very least for Susan. Rather than being treated as a pariah or an obstructionist, she could have been recognized as a key contributor to important work, with significant positive implications for both her science and her career.

The astute reader may have recognized some similarities between this case and the well-known historical account of Rosalind Franklin and the structure of DNA, from whose story this case has been created. Replace Ed with J.T. Randall, head of King's College biophysics unit, Athan with Maurice Wilkins, and Mohan and Ralph with James Watson and Francis Crick and you have what for many is a familiar story (Watson 1968; Sayre 1975; Maddox 2002).

As illustrated by our alternative scenarios, if any or all of the protagonists in this story had had more highly developed self-awareness and interpersonal savvy, Franklin's involvement and recognition might have been very different. In our revisionist version, Randall managed to face his own unwillingness (here imagined, but not inconsistent with the historical facts) to deal with controversy and conflict in a way that might have helped Franklin face and manage her own frustrations. He might have also helped Franklin to recognize the effects of her behavior and act in ways that advanced the science as well as her own career at the same time.

Recognizing Franklin's role in the outcome of this case is in no way meant to excuse any mistreatment Franklin may have endured, either by the various participants or historically. But human interactions are always of a reciprocal nature. We can only guess how the outcome might have differed if Franklin had had an ally such as the hypothetical Randall in the modified case.

Whereas Franklin's data helped reveal the structure of DNA, Susan's data helped reveal a promising cancer therapy. In both cases, the discoveries or breakthroughs almost certainly would have occurred eventually. In the cancer story, the principal tragedy was that an important and possibly lifesaving discovery was delayed. In the DNA story, an excellent scientist who could have been a full collaborator and participant in the discovery of a lifetime was marginalized by Wilkins, Watson, and Crick because of their inability to interact productively with her, and her with them.

Our alternative version suggests that it did not have to turn out this way. Scientists can learn to deal with anger and frustration without damaging their professional interactions, mentors to recognize and address conflict before it becomes explosive, and colleagues to develop collaborations that fairly satisfy the interests of all participants. None of this comes naturally to anyone, least of all to those whose professional lives are spent focused on concepts, data, and technical matters. Just as science can and is learned by those who are so inclined, so too can self-awareness and interpersonal skills. Just as science is taught and required as part of the education of a science professional, so too can self-awareness and interpersonal skills.

TRANSFORMING SCIENCE ORGANIZATIONS

Technical professionals in training are hungry for skills that will increase their effectiveness and success. Educators and funding agencies are becoming increasingly aware of the need for this sort of training. But the challenges to making such changes are many. Let us examine a few.

How do science organizations differ from other organizations?

In previous chapters, we focused on helping individual scientists and managers deal with or overcome challenges in the workplace. Many of these challenges are either unique to or very prominent in scientific organizations. Similarly, science organizations themselves hold challenges that must be overcome or dealt with if they are to function more efficiently. In the following, we review two classes of challenges that face the science organization: those that are unique to science-based organizations and those that are shared with other types of organizations.

It is dangerous to assume that scientists and science organizations can be managed in the same way as other professionals and other organizations. It is equally dangerous to assume that because of their profession, scientists are devoid of the same blind spots and frailties as the rest of humanity. This latter belief can lead to the mistaken notion that science organizations operate with a degree of rationality that is absent from other organizations.

Here is our list of challenges that distinguish science organizations, especially those focused on discovery, from other types of organizations.

- Science organizations engaging in discovery research present unique managerial challenges. Discoveries cannot be made according to a predetermined timetable. This creates an inherent tension between scientists and managers. Managers must produce budgets and time lines based on expectations of when certain events are most likely to occur. Because of the impossibility of scheduling discoveries, a rigid bureaucratic approach to management will result in a disjunction between what scientists produce and what management espouses that they should or will produce.

- Because of scientists' disinclination to deal with interpersonal issues, "people" problems tend to be ignored and, as a result, escalate. Problems that could have been resolved by line or lab managers often end up in human resources, or the chairman's or vice president's office.

- Questioning and skepticism play a central role in the scientific process. However, skepticism expressed by scientists with poor interpersonal skills leads to unproductive and divisive confrontations, reducing productivity.

- Scientists are trained in an environment in which rewards and reinforcement derive from individual achievement. Despite declarations to the contrary, real rewards, promotions, and advancement in any organization are based on individual performance. Cynicism results when scientists are asked to behave as team members by sharing knowledge, insights, and results but observe rewards going solely to those who impress management with individual achievements.

- Science organizations are often led by executives who have little or no science training. In fact, it is not uncommon to find individuals with financial, legal, and business development backgrounds heading such organizations. Such leaders suffer from definite disadvantages if they do not understand the nature of the scientific process, and especially if they believe that research and discovery can be managed by conventional "command and control" approaches.

- During the past two decades, companies in many sectors of science and technology have suffered from financial instability, short lifetimes, and major organizational changes such as mergers and acquisitions. As a result, the stability required to train employees and create a corporate culture is frequently missing. Model programs for organizational change, team building, and management that have been developed based on Fortune 500 companies (as most have) need to be carefully examined for relevance, especially regarding the science sector.

How are science organizations similar to other organizations?

Many of the unique challenges facing science organizations are similar to those faced by other organizations. Nonscientists in particular may believe that because of the nature of science and scientists, such challenges are either absent or unimportant. This is a mistake. The following is a list of challenges that scientists and science organizations have in common with their nonscience counterparts.

- Like professionals in other fields, scientists have a strong need to take ownership of projects and ideas as a way of defining themselves. Managing scientists without understanding this produces a frustrated and unmotivated workforce.

- Scientists have as much of a tendency as anyone else to rely on what they know. It is dangerous to believe that because someone is a scientist, she will dispassionately evaluate all possible approaches or alternatives. When I asked a pharmaceutical company manager why a division in his company was using *Drosophila* (fruit flies) as the model organism for a certain study, he replied, "Oh, that group is run by Dr. Hess. He did his last postdoc in a fly lab."

- Every scientist has a ready willingness to believe that their latest discovery is the most important breakthrough of the century.

- Eminent scientists join advisory boards of science companies for dozens of reasons. A belief in the company's technology may be among these reasons, but not necessarily. Do not judge the value of a company's science, or the likelihood that the company will succeed, based on the number of Nobel laureates on its scientific advisory board. As we demonstrate in this book, science-based companies need more than a stellar scientific advisory board to succeed.

Focusing on people

Science and technical professionals can begin transforming the organizations in which they work, from the ground up, by taking the following steps.

i. Hire the right people

While an executive in biotechnology companies, I divided the qualifications of candidates for scientist and science management positions into three categories: (1) academic or technical skills, (2) professional experience, and (3) personal characteristics and interpersonal skills. The first two categories could be captured on a resume and were typically the basis for the first round of selection. During a candidate's interview, however, I was most interested in the skills in the third category; these skills could not be captured on a resume. Here is a list of key qualifications and characteristics I sought in a potential employee:

1. Academic and technical
   - Technical skills, academic degrees, publications

2. Professional
   - Previous experience in similar organizations or jobs
   - Managerial/supervisory experience as appropriate
   - Strategic experience/vision as appropriate

3. Behavioral and psychological
   - Good self-awareness and ability to step back, observe, and comment on own behavior
   - Willingness to take personal ownership for behavior
   - Ability to describe feelings experienced in a difficult situation
   - Openness to constructive criticism and observations about behavior. Absence of defensiveness during such conversations
   - Ability to control actions under tense or stressful conditions
   - Ability to constructively provide feedback to others
   - Ability to read others well enough to sense which people will work well together
   - Absence of need to dominate or belittle
   - Ability to accept ambiguity without seeking quick and possibly inappropriate solutions to difficult problems
   - Ability to deal productively with implied or direct affronts in a group setting
   - Ability to negotiate equitable agreements when parties have differences of opinion

- Understanding of organizational objectives and how to help meet them
- Ability to remain focused and avoid tangents
- Willingness to seek advice of others when stuck
- Willingness to admit ignorance
- Ability to handle disagreements directly but collegially
- Willingness to share knowledge

Note that none of these relates to intelligence, technical proficiency, the amount of literature that the individual has read, or how often they have been published. Although these latter characteristics might identify "high achievers," achievement per se is not my first criterion for choosing a leader, manager, or even a scientist. I wonder how many of the scientists in a typical company meet more than one third of these criteria.

Here is a list of questions based on the above characteristics to ask yourself about potential candidates.

- How self-aware is this person? Can she step back, observe, and comment on her own behavior?

- Does he take personal ownership for his behavior, or does he explain it as being the inevitable consequence of external circumstances?

- Can she describe her feelings or experiences in some difficult situation?

- Is he open to constructive criticism and observations about his behavior? Does he become defensive or uneasy during such conversations, or receptive and curious?

- Can she control her actions under tense or stressful conditions?

- Can he provide constructive feedback?

- Can she read others well enough to sense the best possible project partnerships?

- Does he need to dominate or belittle others?

- Is she comfortable with ambiguity or does she seek quick and possibly inappropriate solutions to difficult problems?

- Can he deal productively with implied or direct affronts in a group?

- Can she negotiate equitable agreements when parties have differences of opinion?

- Does he understand the organizational objectives and how to help meet them?

- Does she tend to go off on tangents or get so bogged down in detail that she cannot see the big picture?

- Will he seek the advice of others when stuck?

- Does she have difficulty admitting ignorance or is she overly confident of what she does know?

- How does he handle disagreement? Does he remain silent? Does he attack?

- How readily does she share her knowledge?

Sometimes I gained insights by asking candidates for anecdotes or stories about previous jobs. In many cases, my insights came not from the actual content but from other information. Here are three ways to get answers to some of the questions listed above.

1. Ask the candidate to describe their last position and the projects on which they worked. One candidate might be obsessively technical and neglect to mention others who worked on the project, whereas another might discuss the science as well as the group and the way in which the project fit into the company's goals. All other things being equal, I would surmise that the second candidate would be a better candidate for a managerial or leadership position.

2. Ask the candidate to describe an especially difficult problem and how they approached it. Listen for evidence of interactions or collaborations: Was input or advice from others sought and welcomed or were projects bogged down due to unwillingness to seek such input or advice? As above, pay attention to the balance between technical detail and contextual, interpersonal, and organizational references.

3. Pose a hypothetical situation for the candidate. One of my favorites is seeing how he or she would deal with a scientist in a group situation who is very smart and capable but who hoards data and is abrasive to others. If you have read this book, you know that there is no single right answer. The kinds of responses you get will tell you much about how the scientist is attuned to the complexities of managing other scientists. One candidate responded by saying that he would give the hypothetical scientist an ultimatum to either "shape up or ship out." Another suggested that he did not see a problem as long as the scientist shared the data with *him*. Neither of these candidates was high on my list of new hires.

Finally, note that only some of the characteristics listed above can be influenced by the organization and how it is structured, but then only to a limited degree. It is true that dysfunctional organizations will at times produce, and certainly exacerbate, unhealthly interpersonal styles, and in a sense, all boats rise or sink with this tide. The effective person becomes less effective, and the ineffective becomes dysfunctional or dangerous. But the fact remains that even in dysfunctional organizations, individuals must make behavioral decisions and interact with others in ways that will affect outcomes either positively or negatively.

ii. Train the employees that you hire

I know that scientists who have more than a few of the qualities identified by the questions above are few and far between. Consequently, the most effective way to transform your organization is to train employees to be better managers and colleagues. Of course, this is easier said than done. By far, the best way for scientists to learn the skills presented in this book is to observe and emulate those who already have them. If you have these skills yourself, model them for your lab, team, or company. You may wish to take advantage of specific workshops in conflict resolution and negotiation, as well as individual coaching (see Other Resources on p. 171 of the Appendix). Unfortunately, few groups or individuals specialize in providing such training in a way that addresses the specific needs of scientists.

Generic courses in negotiation, teamwork, and leadership can fill some of this void. As noted above, the Appendix lists readings and resources for educating yourself and your scientists. Many of the listed books influenced this book, but none covers the range of topics that we have covered here or focuses specifically on the needs of science and technical professionals.

The exercises at the end of our chapters can be used by individuals or groups as "continuing education" material. But if you choose to discuss some of these issues in a group format, do so with caution. Many people are uncomfortable or embarrassed to discuss themselves, their sensitivities, and their feelings. Never force participants to engage in a discussion if you sense that this may be the case. **Forcing someone with poor self-awareness into a discussion of herself in the presence of others can be at best uncomfortable for that person, and at worst, traumatic, especially if you are not trained in group facilitation.**

Finally, remember that for some, interpersonal skills are far more difficult to learn than technical skills. Unlike technical skills, which are often learned anew, interpersonal skills, or lack thereof, are often deeply ingrained in one's personality. Thus, depending on the individual, there may be a significant amount of unlearning that needs to take place before new learning can be effective. This is no mean feat for most and may be impossible for some.

A manager or leader has only so much time to devote to training scientists in these skills and must judge who is trainable and who is not. Answering the questions in the previous section is one way to sense someone's potential. Training someone with limited self-awareness and little sign of being able to acquire it is not the best use of a manager's time. If you supervise such employees who are valuable from a technical perspective, you may need to find a way to take advantage of their technical skills while ensuring that their interpersonal deficits do as little damage as possible. The case study involving Raju in Chapter 2 is one way to accomplish this. The next section provides a framework within which you can identify those in your organization who, with mentoring, have the greatest potential to become good managers and leaders.

Myths of Scientific Research

- Research is done in mental and social isolation.
- The most important determinant of scientific success is how much science you know.
- Success in science requires a single-minded devotion to science and nothing else.
- Hire the most technically qualified people and do not worry about anything else.

iii. Identify and mentor your future leaders

The single most important element of a transformed science organization is its personnel. Once you have begun training your science staff in the skills of self-awareness and interpersonal relations, you can focus on identifying the leaders. Thus, just as you can identify scientists who have or can acquire the requisite technical skills, so too can you select the ones with the greatest promise to become the future managers and leaders of your organization.

We have found it useful to identify future leaders by assigning scientists to one of three categories.

Category 1. Scientists in this category are technically proficient and already have good interpersonal skills. A subset of these will also have good self-awareness, and the remainder may be able to improve this skill with a modest amount of mentoring. Category 1 is likely the smallest group of scientists in your organization, perhaps fewer than 5%. These scientists need only a modest amount of your mentoring time, largely to help them attain the technical skills needed to manage and continue to develop their self-awareness and relational skills. Give these scientists managerial or leadership responsibilities early in their career so that they can benefit from feedback as soon as possible.

For some readers, identifying Category 1 scientists will be intuitive. They are the ones who interact smoothly with others, take feedback in stride, and avoid getting involved in laboratory cliques or in power struggles. These scientists can also be identified by answering the evaluation questions listed earlier in the chapter.

Category 2. These scientists are technically proficient, have poor or indeterminate interpersonal skills, but sufficient self-awareness to be candidates for improvement. Although they may at times behave in an inconsiderate or insensitive manner, they are nonetheless open to discussion about their behavior. This group may be the one with which the savvy manager can have the most impact. If a scientist is open to observing his behavior and reflecting on his underlying motivations, there is an excellent chance that his behavior can be improved.

One way to identify scientists in this group is to give them some feedback and observe how they react. Denial, defensiveness, and excuses are all red flags. On another occasion you might ask why they behaved in a way that was counterproductive in a group setting, or insulting or devaluing to someone. Gently probe for an indication that they are willing to examine their motivation and can understand and empathize with the victim of their disfavor. Positive signs in either of these circumstances indicate that you may continue such discussions, and help the employee use her self-awareness to gain better control of her behavior and improve rapport with others. Helping a scientist in this category can be one of the most satisfying things you do as a manager or leader, and can have the greatest impact on your organization.

Category 3 contains scientists who are technically proficient but have poor interpersonal skills, low self-awareness, and a limited capability to improve or limited interest in improving. This is usually the largest of the three groups. These employees

- Become defensive when observations are made about their behavior or performance

- Find it difficult or impossible to discuss their feelings or motivations

- Have a history of being confrontational, argumentative, and devaluing of others, or

- Have a history of being passive, withdrawn, and uncommunicative

Scientists in this category are often the most knowledgeable, technically adept, and scientifically productive. But they are unlikely to see the utility of the skills offered in this book, and they may be resistant to efforts aimed at helping them to acquire such skills.

If Category 3 scientists have superior, unique, or essential technical or scientific skills, place them in a position from which they can do the most good scientifically and the least damage interpersonally. The example of Raju in Chapter 2 shows one way in which such a position can be created for a senior individual.

For valuable intermediate-level scientists, you may be able to find ways to incorporate them into teams that allow them some flexibility. Often, insisting that such people work in a highly structured or interactive team setting is a recipe for disaster. Making them responsible for specific tasks or deliverables that can be accomplished independently or with limited group interaction is one way to solve this problem. The key is to maintain a formal relationship between the scientist and the team, with well-defined guidelines and responsibilities.

Often, those in this category are very concrete thinkers who function best with clear and unambiguous assignments. Be certain that the responsibilities and deliverables that you assign are defined in as clear a manner as possible. It goes without saying that you should think twice before giving such employees line management responsibilities. If you must place them in a managerial position or have others reporting to them, be alert to signs of stress in those individuals, and be prepared to intervene.

There are always exceptions. We have seen cases where scientists who were viewed as having weak interpersonal and managerial skills managed to be effective team leaders. In one case, a senior scientist, who under most circumstances would never intentionally have been given managerial responsibility, had attracted, over time, a small group of dedicated scientists very much like himself in personality and temperament. This group was remarkably effective and productive. Although they worked in virtual isolation, their role was such that this was not a major detriment. Their work output was integrated into those projects that required it, and feedback was provided though ad hoc but effective communications. This kind of group would be almost impossible to create intentionally, but if you have one in your organization, think carefully before you disband it.

TRANSFORMED SCIENCE ORGANIZATIONS

Transformed science organizations consist of employees and leaders who share the belief that the way in which scientists do science has an impact on the science that they do. Such organizations

- Consist of members who think about and manage how they interact with one another, deal with conflicts productively, and iron out animosities before they become feuds.

- Carry out both internal and external negotiations respectfully, directly, and honestly.

- Have leaders who pay attention to their own behavior and interactions, and encourage others to do the same.

None of this is to suggest that transformed science organizations are populated by scientists with saccharine smiles behaving like lobotomized drones. A transformed science organization does not eliminate the spark that ignites creativity, or the healthy disagreements that often stimulate discovery. Rather, it provides science and technical professionals with tools to deal with the inevitable nonscientific problems of the scientific workplace. These tools free them to direct all of their energy and excitement toward solving scientific problems and accomplishing the organization's objectives.

The underlying reactions that ignite nonproductive behaviors are universal. They cannot, and probably should not, be extinguished. Human feelings of jealousy, anger, fear, competition, and more are natural. Those in groups will, at times, feel ostracized, misunderstood, and isolated.

Transforming science organizations does not make these emotional reactions disappear. Rather, it provides scientists with tools to behave effectively in response to these reactions. It also provides them with a repertoire of skills on which they can call to resolve those issues that lead to strong reactions.

Transforming a science organization cannot be accomplished without the support and attention of its most senior scientific leaders. "People" skills must be taught to students, trainees, and employees in the same way that technical skills are taught. We must stop referring to such skills as "soft" and stop characterizing those who espouse them as "touchy-feely." Moreover, we must either stop relegating the training for such skills to human resource departments, or imbue these departments with an importance and focus that they rarely have in most organizations. This will not happen unless we agree to classify these skills as "must-have" and not "nice-to-have."

We have identified the several categories of skills that should be taught to scientists. First are skills that help scientists to become better aware of their own behavior. These skills are prerequisites for productive interactions under difficult or ambiguous circumstances. The second set of skills helps scientists to become better attuned to others and how others respond to them. Third are skills that help scientists to become good negotiators and to reach agreements that are equitable and satisfy the interests of all parties. We showed how such skills can improve the way scientists interact with peers, supervisors, and team members. We showed that when trainees and mentors applied these skills, it is possible to expand the scope of science education and create scientists who are as knowledgeable about how science is done as they are about the science itself. We also outlined what scientists need to know and the skills that they need to use to thrive in the private sector. Finally, we suggested ways in which science organizations can use our recommendations to select the best candidates for scientific and leadership positions.

The skills and tools we presented can be taught through workshops, instructional materials, and case studies. As we noted, we believe that workshops are invaluable for teaching negotiation and associated skills because of the immediacy of the learning experience. Case studies such as the one at the beginning of this chapter (see p. 150) can be used to present complex science-management problems. They can also be used to show how scientific progress is affected by the quality of the interactions of the people involved. These studies may be used as didactic tools on their own or to catalyze discussion in formal training settings.

We believe that with research and experimentation, we can develop additional approaches to help scientists improve communications, interactions, and collaborations. This book offers a starting point. But vision and support is needed to develop other options and approaches. We call on educational institutions as well as agencies and foundations that support scientific research to take the lead.

REFERENCES

Maddox B. 2002. *Rosalind Franklin: The Dark Lady of DNA*. Harper-Collins, New York.
Sayre A. 1975. *Rosalind Franklin and DNA*. W.W. Norton, New York.
Watson J.D. 1968. *The Double Helix: A Personal Account of the Discovery of the Structure of DNA*. Atheneum, New York.

EXERCISES AND EXPERIMENTS

1 Identifying potential leaders

Becoming aware of who among your scientific staff are good candidates for leadership and managerial responsibility will enable you to mentor and advise your leaders-in-training. Answering the following questions can help you to identify candidates for Category 1 or 3 (for a description of scientists in Categories 1–3, see pp. 162–164). Enter a number for each person you are evaluating in the column to the right of the trait listed: Enter a 5 if you strongly agree with the statement as it applies to the person and a smaller number if you agree less. Add them up for a measure of your views of those you are rating. The higher the number, the better the potential of the candidate. This exercise is not meant to be used to rank your staff in any absolute way, but rather as a tool to help you ask the appropriate questions when you are considering the leadership potential of your staff. Remember that the rankings are your own views, which may themselves be skewed or biased.

	Strongly agree 5	4	3	Disagree 2	1
1. Person can step back, observe, and comment on her own behavior.					
2. When prompted to observe and comment, is open and willing.					
3. Takes personal ownership for behavior.					
4. Does not seek to blame others or external circumstances for mistakes or failings.					
5. Able to describe feelings or experiences in difficult situations.					
6. Open to constructive criticism and observations.					
7. Does not become defensive or uneasy during such conversations.					
8. Able to control behavior under tense or stressful circumstances.					
9. Can deal with implied or direct affronts in a group.					
10. Can provide consructive feedback.					
11. Does not dominate or belittle others.					
12. Able to negotiate and arrive at equitable agreement when parties have differences of opinion.					
13. Understands organizational objectives and how to help meet them.					
14. Can see the big picture associated with a project.					
15. Willingly seeks the advice of others.					
16. Has little difficulty admitting ignorance.					
17. Is not overly confident of own knowledge.					
18. Does not avoid or withdraw from disagreements or difficult conversations.					
19. Willing to share knowledge.					

Of these 19 items, traits 1, 2, 5, 6, and 7 relate to the ability of the individual to listen and respond productively to observations about their behavior and to have insight into it. Low scores in these traits should not be viewed as having absolute psychological validity. These are your assessments of the person's ability to be open and receptive to discussing their own behavior and motivations with you. If your evaluations for each of these traits are consistently low, then you may have difficulty establishing sufficient rapport to help these individuals observe and improve their behavior. **Because others in your organization may have a better rapport with specific individuals than you, always validate your perceptions about leadership or management potential with one or more of your colleagues or a contact in human resources.**

2 Rewriting your own history

Review the case study in this chapter. Consider a situation from your own organization or group, in which you were involved, that had a negative impact on a project or the individuals.

- Summarize the incident, focusing on the impact on a project, task, or objective, and the individuals involved, as well as the consequences for the team or organization.

- List as many ways as possible that would have resulted in the incident turning out differently or mitigating the impact.

- Refer to the skills presented in this book. Choose one skill and consider how acquiring or exercising that skill by one of the participants in the incident might have changed the outcome for the better. Try this for as many of the individual skills and participants as possible.

- Focus on your own involvement in the incident, and do the same for yourself as you did above for the others. Identify as many ways as possible that exercising the skills could have helped you to effect a better outcome.

- Use the results of this exercise to anticipate ways in which you can avert or mitigate future incidents that may impact your work.

Appendix

RESOURCES

Books

At the Helm: A Laboratory Navigator. Kathy Barker (2002), Cold Spring Harbor Laboratory Press, Cold Spring Harbor, New York.
A "must have" for every new manager of a science group. Written by a working scientist, provides concrete and useful answers to questions about organizing and administering a scientific laboratory and group. Topics also include psychological issues such as lab culture, conflict, and morale.

Dealing with People You Can't Stand: How to Bring out the Best in People at Their Worst. Rick Brinkman and Rick Kirschner (1994), McGraw-Hill, New York.
Builds on some of the themes developed in *The Feeling Good Handbook* (see below) and provides extended illustrations and amusing examples.

Developing Managerial Skills in Engineers and Scientists: Succeeding as a Technical Manager. M.K. Badawy (1982), Van Nostrand Reinhold, New York.
Provides guidance on choosing science management as a career and how to prepare for it. One of the few books included here that contains material on planning, decision making, and project management.

Emotional Intelligence. Daniel Goleman (1995), Bantam Books, New York.
Excellent guide to this important topic. Illustrates nature of emotional intelligence and argues persuasively for its importance in our daily lives and interactions.

The Feeling Good Handbook. David D. Burns (1990), Plume/Penguin, New York.
Excellent and highly readable introduction to the concepts and applications of cognitive behavior therapy, on which our approach to self-management is based. An outstanding book for those seeking further guidance on dealing with difficult people (see Chapters 20–22). Our use of the "agree, empathize, inquire" approach in Chapter 5 is taken from these chapters of Burns' book. Also contains useful insights into understanding and managing your moods and interactions.

Getting Past No: Negotiating Your Way from Confrontation to Cooperation. William Ury (1993), Bantam Books, New York.

This short paperback is probably the single most useful book written on the topic of negotiation and is the foundation of Chapter 3 of our book. Should be required reading for every scientist, regardless of whether they aspire to be a manager.

How to Manage the R&D Staff: A Looking Glass World. James E. Tingstad (1991), Amacom, New York.

Written primarily for the experienced manager, provides guidance on many of the practical aspects of running a research laboratory from a psychological perspective. Covers such topics as group cohesion, building self-confidence in scientists, and task delegation.

The Human Side of Managing Technological Innovation: A Collection of Readings, 2nd edition. Ralph Katz, Ed. (2004). Oxford University Press, New York.

Fascinating and wide-ranging collection of readings on managing in science and technology organizations. Topics include teamwork, creativity, and managing scientists.

Leadership Without Easy Answers. Ronald A. Heifetz (1994), Belknap Press/Harvard University Press, Cambridge, Massachusetts.

An insightful study of what leaders do and why. Focuses on why those in leadership positions feel pressured to provide quick and easy answers to complex issues, and how to recognize and avoid this. Introduces the important concept of "managing ambiguity." Contains many useful lessons for those who lead science groups.

Leading Geeks: How To Manage and Lead People Who Deliver Technology. Paul Glen (2003), Jossey-Bass, San Francisco, California.

Insightful book about managing technical professionals. Although it gives the impression of having been written primarily for the manager with a nontechnical background, it is full of insights and ideas that scientists will find enlightening.

Managing Scientists: Leadership Strategies in Scientific Research, 2nd edition. Alice M. Sapienza (2004), Wiley-Liss, Hoboken, New Jersey.

An excellent introduction to various approaches that may be taken by scientists who manage other scientists. Addresses psychological and cultural issues in ways that are very accessible to most scientists and science managers.

Organizational Culture and Leadership. Edgar H. Schein (1992), Jossey-Bass, San Francisco, California.

Classic study of organizations and organizational culture. Despite its rather academic-sounding title, contains case studies and insights that are illuminating for scientists seeking to understand the psychological underpinnings of the behavior of those in organizations and groups.

Scientists in Organizations: Productive Climates for Research and Development. Donald C. Pelz and Frank M. Andrews (1976), Institute for Social Research, University of Michigan, Ann Arbor.

Somewhat dated but interesting use of quantitative techniques to uncover the organizational and individual correlates of creativity in scientists and managers of science.

Other Resources

The A.K. Rice Institute for the Study of Social Systems
(http://www.uvm.edu/~mkessler/akrice/)

A.K. Rice Institute, through its local affiliates, sponsors experiential workshops that focus on providing insight into the way in which people interact and work in groups. The Institute "...seeks to deepen the understanding and the analysis of complex systemic psychodynamic and covert processes which give rise to nonrational behavior in individuals, groups, organizations, communities and nations." Do not be put off by the jargon. We recommend these workshops for those who are interested in and committed to learning about their own behavior and reactions in groups in a work setting.

American Group Psychotherapy Association (AGPA)
(http://ww.AGPA.org)

A professional organization that trains human services professionals in group therapy and group dynamics. Also provides a national referral service for those seeking certified group psychotherapists. Participation in group psychotherapy is an excellent way to acquire emotional intelligence skills for those who have difficulty learning these skills on their own.

Emotional Intelligence (EI)

Dozens of books have been written on this topic, including applications of EI to the workplace and leadership development (see, e.g., *The Emotionally Intelligent Manager*, by David R. Caruso and Peter Salovey [2004], Jossey-Bass, San Francisco, California; *Primal Leadership: Realizing the Power of Emotional Intelligence*, by Daniel Goleman, Richard E. Boyatzis, and Annie McKee [2002], Harvard Business School Press, Boston, Massachusetts). Books on improving your EI are also available (see, e.g., *Raising Your Emotional Intelligence: A Practical Guide*, by Jeanne Segal [1997], Henry Holt, New York; *The Emotional Intelligence Activity Book: 50 Activities for Developing EQ at Work*, by Adele B. Lynn [2002], Amacom, HRD Press, New York). In addition, a growing number of online resources offer information, assessment quizzes, and links on emotional intelligence. One reputable commercial organization with which we are familiar is the Hay Group (http://ei.haygroup.com). A Google search done in September 2004 returned over 1,000,000 hits to the inquiry "emotional intelligence." *Caveat emptor.*

National Training Labs (NTL)
(http://www.ntl.org/about.html)

A highly regarded nonprofit organization that originated the "T-group" concept for experiential learning. NTL specializes in experiential training in leadership, organizational, professional, and personal development. Readers of our book may be particularly interested in their "Human Interaction Laboratory": "...an introduction to interpersonal relations, group dynamics, with a focus on developing and practicing effective interpersonal skills and giving and receiving feedback responsibly."

The Program on Negotiation of the Harvard Law School
(https://www.pon.harvard.edu/main/home/index.php3)

Offers numerous lectures and workshops throughout the year focused on education, techniques, and applications of negotiation and mediation. Web site contains links to a variety of resources, seminars, courses, and materials that the program makes available for purchase.

Science Management Associates
(www.sciencema.com)

Science Management Associates provides workshops, training, and coaching to scientists, science managers, and executives who wish to improve their interpersonal, group, and organizational skills. These workshops are presented throughout the year at scientific meetings and conferences, and can be customized for presentation at individual companies and other scientific organizations. Clients have included "top ten" pharmaceutical and biotechnology companies, and educational institutions.

Index

About the Authors

Carl M. Cohen is Chief Operating Officer of Biovest International and President of Science Management Associates.

Suzanne L. Cohen is a Licensed Psychologist with a private practice in Wellesley, Massachusetts and is a Clinical Instructor in Psychology at Harvard Medical School.